The Human Scaffold

GREAT TRANSFORMATIONS

Craig Calhoun and Nils Gilman, Series Editors

1. *Renovating Democracy: Governing in the Age of Globalization and Digital Capitalism*, by Nathan Gardels and Nicolas Berggruen

2. *The Human Scaffold: How Not to Design Your Way Out of a Climate Crisis*, by Josh Berson

The Human Scaffold

HOW NOT TO DESIGN YOUR WAY
OUT OF A CLIMATE CRISIS

Josh Berson

UNIVERSITY OF CALIFORNIA PRESS

Berggruen
Institute

*The publisher and the University of California Press Foundation
gratefully acknowledge the generous support of the Ralph and
Shirley Shapiro Endowment Fund in Environmental Studies.*

University of California Press
Oakland, California

Library of Congress Cataloging-in-Publication Data

Names: Berson, Josh, author.
Title: The human scaffold : how not to design your way out of a climate
crisis / Josh Berson.
Other titles: Great transformations ; 2.
Description: Oakland, California : University of California Press, [2021]
| Series: Great transformations ; 2 | Includes bibliographical references
and index.
Identifiers: LCCN 2020037015 (print) | LCCN 2020037016 (ebook) |
ISBN 9780520380486 (cloth) | ISBN 9780520380493 (paperback) |
ISBN 9780520380509 (epub)
Subjects: LCSH: Climatic changes—Social aspects. | Climatic changes—
Effect of human beings on.
Classification: LCC QC903 .B477 2021 (print) | LCC QC903 (ebook) |
DDC 304.2/5—dc23
LC record available at https://lccn.loc.gov/2020037015
LC ebook record available at https://lccn.loc.gov/2020037016

Manufactured in the United States of America

29 28 27 26 25 24 23 22 21
10 9 8 7 6 5 4 3 2 1

For Jessy

He decides that after this the only farms and roads that he can safely build will be tiny lumps and faint roads so absurdly small that even he, their designer, will have to believe that he sees them from across an enormous distance, and even wonders whether he should make his backyard the country of a people like the Aborigines or even some earlier race of people who made no marks at all on the grasslands or in the forests so that he can follow their journeys without plucking out a single weed or altering the lie of the least patch of dust.

Gerald Murnane, *Tamarisk Row*

Contents

List of Figures		ix
Preface: Living Epiphytically		xi
Kansha		xxxiii
1.	Treadmills	1
2.	Scaffolds	30
3.	Equilibria	55
4.	Landscapes	76
4boro.	Landscapes and Scaffolds	110
5.	Ditch Kit	125
	Postscript: Foaminess	155
	Glossary	169
	Notes	175
	Sources	193
	Index	209

Figures

1. Popup studio, 1898. Members of the Cambridge Anthropological Expedition to Torres Straits — xxix

2. Bark-bundle canoes, eastern shore of Schouten Island, 1802 — 19

3. Haenyeo warming themselves at a *bulteok* (enclosed hearth), Jeju Island, 2004 or 2005 — 50

4. Successive pulses from CP 1919 (now known as PSR B1919+21), 1968 — 74

5. The Gray-Scott reaction-diffusion model — 108

6. Two ways of imagining evolution: networks of milieux and scaffolds — 123

7. Games with stuff — 141

8. Skills reservoirs and the survival of "strictly dominated" strategies — 160

9. The multiple concurrent time horizons of niche construction — 164

Preface

What do you do? Where do you live?

Common questions, questions, I imagine, most readers of this book have had cause to ask and answer, perhaps more often than they can recall. For me they pose a challenge. This book is, obliquely, about why I find it so difficult to say what I do and where I live. Partly for this reason, it has been remarkably difficult to write. This is not something I understood at the time. As I wrote the chapters that follow, I ascribed the distinctly effortful quality of my writing days to the cumulative fatigue of having written, depending on how you count, three, or four, or five books back-to-back, to an unfamiliar rhythm of professional responsibilities that obtruded into my working week, to the strain of writing and managing these new responsibilities while at the same time setting up a house in an unfamiliar city—Los Angeles—after ten years living outside the United States, and to my shock—common, I discovered, among the newly returned—at how the country had changed in those years. These all played a role. But mainly, I have come to see, it was how the themes of this book touched a distinctly personal nerve that made writing it so difficult. Even now, this morning, writing what should be the easy part, I feel an awkwardness, I find myself straining to hear the music, as if from a neighboring room with a closed door between and a kettle coming to a boil at my elbow—testament, no doubt, to how awkward I find writing about myself.

In fact, I did not intend to include a preface at all. On the morning at the end of August, just a bit over ten months back, when I started writing this book in earnest—not proposals or sketches or notes but what I understood would form the published text itself—I started with the following:

A Note to the Reader:
Writing this book, I've tried to put myself in a contemplative frame of mind. I encourage you to do the same when reading it. Accordingly, there is no prefatory material. The book begins *in medias res* and its themes emerge organically. If you find yourself desperate for a more explicit delineation of theme, you have my blessing to skip to the final chapter.

This was a bit dishonest, because of course "writing this book," the experience whose outcome, by implication, informed the choices— "Accordingly, there is no prefatory material"—and illocutionary acts— "I encourage you . . . you have my blessing"—lay entirely in prospect. When I wrote this paragraph its content was aspirational. What I had, that morning, was not a textual basis for dispensing with the signposting and, to use a word that will recur, the scaffolding typical of a research essay, but a desire to write something freestanding and self-contained, something that, without sacrificing the rigor and precision that I prize above all else, would not tax the reader in the way, I had come to see, my previous efforts had done. Something not glib but accessible, something that disclosed itself the way the wall discloses itself when you practice zazen early in the morning, the texture and scuff marks, the movements of insects, the text on the spines of books, if you are facing bookshelves, filling in as the hour unfolds and the dark gives way to the gray wash of an overcast sky—or the pale yellow of sunrise—as you sit, hands folded atop the medial process of the calcaneal tuberosity—I tend to start out with the right foot supported by the left, switching halfway through—thumbs forming a bridge, breathing slowly, steadying the gaze. In a way this too is dishonest, for when I wrote "A Note to the Reader" I was not aware that what I had in mind by "a contemplative frame of mind" was something so specific, though I am confident now, for reasons made clear in chapter 4, that it was.

Mainly, I wanted the reader not to have to work so hard. And I wanted not to have to work so hard myself.

As I write, the house is a mess. Two days ago we returned, my partner and I, to our place in Berlin—Jessy's place, really, her home for the past fourteen years, mine, on and increasingly off, for four—after ten months (for Jessy) or eleven (for me) away. Our subtenant did a reasonable job keeping the place intact. Still, the floors feel gritty and the surfaces are covered with nests of varied debris: T-shirts to be washed, packets of hempseed powder and cacao nibs, charging cables documenting the evolution of serial bus standards over the past ten years, environmentally friendly vessels and utensils in stainless steel, titanium, and bamboo, a letter from the building management indicating that the rent will go up 15 percent at the start of September. In the fridge are unfamiliar containers of things we would never keep around. The craquelure on the surface of the bathroom sink has grown. A new washer stands in the kitchen, between hob and sink, where we had it placed seven months ago, arranging the whole thing from Los Angeles.

This book, for all that it is brief and, to my way of thinking, dissatisfactory, has been close to eight years in the making. It was October or November 2011 when I first came across the materials that form the basis for chapters 1 and 2. It was April 2014, not long before I met Jessy, when I first started making notes about the peripatetic character my life had taken on, my difficulty saying where I lived, or even where I was "based." But it is really in the past two years that this book has taken form, and the places where I conceived and wrote it speak to its themes. These included a one-room cabin on Lough Derg, Ireland, in June 2017; a skeuomorphic shepherd's hut on the Isle of Eigg, in the Inner Hebrides, in July 2018; a trailer on a subdivided ranch in Antelope Valley, on the southern rim of the Mojave Desert, in November 2018; and, most significant of all, a cabin in Onyuudani, a remote hamlet in the Lake Biwa watershed, on the outskirts of Takashima, Shiga prefecture, north of Kyoto. With the exception of this preface and brief sections at the ends of chapters 4 and 5, I wrote the text itself in a backyard cottage in the shadow of the ridgeline separating the Highland Park and Mount Washington districts of Los Angeles. On satellite images, the house appears to sit at the edge of a large park, but in fact this is a hill so steep, and so thick with coarse dryland vegetation, as to be nearly unnavigable. You could, if you wanted, hike up to the ridgeline through the notional park, but most days you were better off taking the long way around. In any event, the fact that the house stood in shadow most of the day, especially—as the ridge stood to the west—in the afternoon, meant that it tended to be two or three degrees cooler in our home than out in the main road. In August, when I arrived, this was a blessing. In winter it made writing a challenge—as with many small structures in winterwet climates, our cottage was characterized by a distinct absence of insulation—but a productive one, as thermoregulation has come to play a prominent role in the argument that follows.

I could name other places that influenced this book. A one-room cabin—styled a bothy though it was lightly built of modern materials and not really on the way anywhere—where we spent a couple nights, on a farm in Inshriach, in the Scottish Highlands, in September 2016. A cool plastered house facing a stand of eucalyptus, with the scent of the ocean, in the village of Odeceixe, in the Algarve region of Portugal, where we finished a four-day hike in September 2015. Like all books, this book has a perspective. One way to think of the perspective this book offers is that of the knapsack, which has become a metonym for my way of being in the world. In the cinematic way that some of us, myself included, have of reflecting on our lives, I imagine the knapsack as a participant in a shot/reverse shot: first you see the knapsack, sitting on the floor, its drybag closure lending it a fig-shaped aspect, then you see the bare room as if from the knapsack's point of view. But lately I have come to think of the perspective this book offers in a different way: this is a book about living epiphytically.

When I moved to Los Angeles eleven months ago I took four books with me. One was Sylvia Hallam's *Fire and Hearth* (Australian Institute of Aboriginal Studies, 1975), discussed in chapter 3. Another was Daniel Friedman and Barry Sinervo's *Evolutionary Games in Natural, Social, and Virtual Worlds* (Oxford University Press, 2016)—evolutionary game theory lurks in the background through much of this text. Then there were two books on plants: Hamlyn Jones's *Plants and Microclimate: A Quantitative Approach to Environmental Plant Physiology* (Cambridge University Press, 2014), and Kathy Willis and Jennifer McElwain's *The Evolution of Plants* (Oxford University Press, 2013). The Friedman and Sinervo and the Jones texts did not make it back from Los Angeles—there is a continuing work of selection, sifting, sloughing off, that unfolds when you move around a lot, and it is one of the things that I find exhilarating about living as I do. *Fire and Hearth*, I suspect, may be with me for some time, if only because it is difficult to find a copy and as a finding aid for a large body of primary sources on the role of fire in winterwet foraging communities it has not really been improved upon in forty-five years—and it does not take up much space. *The Evolution of Plants* is with us now mainly because Jessy has been reading it. But seeing it on the shelf the other day, amid the flotsam of recent arrival, I was reminded of how keen I was, a year ago, to bone up on plant ecology, how urgent this felt—I could not, I felt, do justice to the questions of niche construction, in particular the human manipulation of vegetative cover, that occupy the first

60 percent of this book, absent a firmer grounding in the ecology of plants. I still feel this was reasonable. But now I see something else at work in my turn to the plant world: I was looking for a metaphor for the strategies of survival described in this book.

Our home in Los Angeles was filled with epiphytes. Actually, it would be a mischaracterization to say it was filled with anything. I seem to have a deep-rooted distrust of furniture, of stuff, but I do like having plants around. To mark Jessy's arrival in Los Angeles in October, I ordered a pair of *myouga* seedlings (茗荷, *Zingiber mioga*) from a nursery in Oregon. We planted them, with for me uncharacteristic optimism, in a pair of planters in the yard where a yearslong drought had seen off jasmine, agave, aloe, and other plants far more tolerant of a dry climate than *Z. mioga*, better suited to the monsoonal climate and deciduous forests of southern Japan. One of the first places we went together in Los Angeles was to the Huntington Gardens in Pasadena. By chance, the day we went for the first time, they were holding an orchid sale. From a tray of discarded orchids, Jessy chose a *Dendrobium*. It had been marked down to five dollars. Within a day or two of getting it home, we understood that this was because it was infected with some kind of virus or microfungus. The pitting and black spotting that indicate mesophyll collapse appeared on many of the leaves. Jessy cut back the worst-affected, sprayed the rest with a dilute solution of white vinegar, and sprinkled baking soda over the rhizomes and at the bases of the leaves. I was skeptical that these measures would have an effect, but the plant made a full recovery, later growing to the point where it had to be propped against the house to keep it from tipping over when we removed it from its weighted outer container to give it water and let it sun itself on the porch.

Our *Dendrobium* flourished in a bed of loose gravel, so it would be more accurate to say it was lithophytic than epiphytic, though as it grew its rhizomes spilled out over the lip of its container, as if probing its environment for something ligneous to grasp on to. But we kept other plants that were true epiphytes in that they could

not be embedded in any kind of mineral matrix, however loose—
they would only grow suspended in air, rhizomes preferably coiled
around some other plant. These, I admit, I found a bit finicky. Set
outside to get some sun, they were forever blowing off the rail, and
you had to be careful not to let them get too much moisture. When
I think of the plants that I found most inspiring in the time I was
writing this book, those that filled me with humility and peace, it is
trees that come to mind—the ghost gums mentioned in chapter 5,
the *Casuarina*, deodar, and Montezuma cypress that we would visit
at the Huntington, the *Melaleuca* along Monte Vista that became
visible to me only after I'd written about paperbark watercraft in
chapters 1 and 2. Perhaps the precariousness of the epiphyte strategy
feels a bit too familiar for me to see in it something worthy of respect.

So far I have said something about my trouble with *Where do you
live?*, nothing about *What do you do?* Here too, I have been something of an epiphyte, socialized, in different places and at different
points in my career, as a historian, philosopher, anthropologist,
computer scientist, cognitive scientist, and design thinker, whatever that may be. I did not set out to become a disciplinary skep-
tic, though on balance I think it has served me well—or at least,
the embrace of disciplinary identity runs contrary to my character.
In the five years before I wrote this book, my main institutional
affiliation, though it was a loose one, was with a functional brain-
imaging group at a cognitive science institute, where I saw it as my
role to goad the PhD students and postdocs toward an appreciation
of the value of ecological validity—how people behave, as it were,
in the wild—as a criterion in the design of imaging studies. These
days, when I am obliged to provide a disciplinary epithet, I usually
refer to myself as an anthropologist, because anthropology in the
broad sense—the study of how culture mediates human adaptation
to environment, with emphasis on the ecological determinants of
behavior, the coevolution of individual, community, and milieu, and
the nonlinear interaction of phenomena unfolding over timescales
of ten milliseconds to one million years—feels like the best fit for my

own methodological aspirations. At the same time, I find myself at odds with anthropology as it is practiced today in either of the going disciplinary camps, the interpretive and the analytic (or the constructivist and the reductivist). The one treats precision and rigor in the description of behavior as suspect principles irreconcilable with epistemological pluralism and respect for diversity of experience, the other pursues precision in a deductive fashion that seems to take dimensionality reduction as an end in itself rather than as a provisional, iterative strategy for making sense of a phenomenon—the behavior of encultured beings—that is intrinsically high-dimensional. Of course these are caricatures, and I am not alone in my desire, as anthropologists Agustín Fuentes and Polly Wiessner have put it, to reintegrate anthropology.

But really, more than one colleague has said to me, *you're trying to create a new discipline*. Indeed, it might be something you could call *sensorimotor ecology*—or, extending the project beyond sensory and motoric behavior in the conventional senses, *semiokinetic ecology*. This is a theme I return to in the postscript. Here I will simply note that in the text that follows I do a lot of switching back and forth between analytic and interpretive registers. This is partly a matter of thematic emphasis and choice of evidence, but it plays out in diction and syntax too. By design, parts of the text are cool, crisp, free of emotional coloring or overt indications of my own opinion on some matter of contention, while other parts are personal and charged. For some time, my friend and colleague the artist Simon Penny has had a project called Orthogonal, the aim of which is to build a prototype for a modern oceangoing proa—an asymmetrical dual-hull sailcraft modeled on those long used in Micronesia. One of the design characteristics of proas is a strategy for catching the wind known as *shunting*, "reversing end for end, with the *ama* [outrigger hull] always on the windward side. . . . . This kind of asymmetry," Penny explains, "presents both opportunities and difficulties. It permits light, fast craft of extremely shallow draft, but shunting the rig traditionally involves dragging sail and yards to the other end of the boat." I have done my best to make the shunting between analytic and interpretive registers smooth and nonkinetic. It is my hope that no reader will find their feet tangled in the yards.

But the shunting serves a purpose. I hope it is clear from what follows that I believe deeply in the emancipatory potential of rigorous observation and that I care deeply about getting things right— contextualizing claims, testing them, exposing their methodological and political assumptions, sifting evidence. That these two principles, emancipation and rigor, should be mutually reinforcing rather than mutually inconsistent has long seemed to me not self-evident but more consistent with the evidence than any other position. But more than one sympathetic early reader has pointed out that this is an uncommon position today and that it warrants commentary, perhaps contextualization of its own. *You're out of step*, people have told me, and it is important that I acknowledge my out-of-stepness at the outset.

I am loath to descend into genealogy, either my own or that of a discipline save, as in chapters 1 and 2, as disciplinary genealogy impinges on questions of method in sensorimotor ecology and on the broader questions of policy and values that these give rise to— above all, in this book, the question *What is to be the role of stuff in our lives?* I will say that if I am out step, my teachers have operated in a similar spirit. It is from the linguistic anthropologist Asif Agha that I have adopted, no doubt clumsily, key parts of the outlook that inform this book, particularly in chapters 1, 2, and 3: my skepticism that the principal social function of coordinate action is the transmission of information in the fashion envisioned by proponents of the trait-transmission theories discussed there, as well as my emphasis, verging at times on self-parody, on the metapragmatic dimension of behavior, the way that everything we do, above all everything we do with others, serves either to bolster or to challenge social norms, and generally both at once. Every gesture we make, be it speech in the conventional sense or some other kind of sensorimotor act, represents a link in a chain of norm-enregistering behavior, a chain that can only be grasped by understanding the community, the matrix in which registers of sensorimotor behavior form and dissolve, as a phenomenon of meaning-making as much as, perhaps more than, the individual is such a phenomenon. I have used the term *enaction* to describe the phenomena of coordinate action I have in mind, in order to emphasize that enregisterment, as I have come to think of it over many years of conversation with Asif, is continuous with meaning in the more "cognitive" sense (that is, concerned with how we construe the information about the world conveyed by our senses, above all information about the interior states of other enminded presences) that has typically preoccupied philosophers of mind.

Enaction, in this usage, refers to how socially coordinate movement gives rise to thinking, broadly construed. In my view, this *gives rise to* is evolutionary and developmental as well as functional: it is not that our presence in the world is given by central nervous phenomena, with sensorimotor faculties serving to support the work of the central nervous system—it is, rather, sensorimotor (or, to adopt an agnostic stance toward the nature and degree to which sensing and movement are endowed with an experiential character, semiokinetic) phenomena that are prior, that stand at the center of what it means to be present in the world, to be an awareness sink. Nervous systems, cognition, sentience, indeed, culture, represent outgrowths of signaling and movement. They arise in response to particular kinds of constraints—the need to coordinate signaling and movement across contingently coupled parts of a living thing, as in the processes of central pattern generation that subserve respiration and digestion, or the need to respond to an environment that changes too rapidly for other signaling strategies, for instance the epigenetic regulation of gene transcription, to keep up.

It is from Riki Kuklick that I have adopted another key part of the outlook that informs this book, roughly, the contextualism that guides my reading of "technical" literature (for instance, that published in peer-reviewed journals), be it in anthropology or any discipline, as primary source material. In one of her last published essays, "Personal Equations," Riki undertook precisely the task I myself, a couple paragraphs back, said I did not have the stomach for: she offered an account of how anthropology developed as a discipline—specifically, one in which the practitioner's bodily suffering in the course of extended fieldwork came to be understood as the key both to the rigorous, empathic observation of culturally distant peoples and to socialization in a self-reproducing community of professionals, that is, a discipline. For those most often credited with the formulation of anthropology as a discipline marked by a distinctive method and theory of knowledge, notably W. H. R. Rivers and Bronisław Malinowski, there was no conflict between the rigorous observation of behavior and the sensorimotor and emotional affinity one developed for one's field interlocutors by virtue of prolonged immersion in the ecological setting where one's interlocutors got their living. Indeed, bodily identification formed the basis for rigorous observation. This is something I explore briefly toward the end of chapter 3. I miss Riki. No doubt this would be a better book had I had the chance to discuss it with her.

It strikes me that both these things—a concern with the metapragmatic dimension of coordinate meaning-making, and a concern with how contingent features of life history, above all the emotionally charged circumstances in which mentorship unfolds, shape one's outlook—represent forms of contextualism. Context is not genealogy—indeed, context, it might be said, is the complement of genealogy, comprising everything other than the vertical chain of transmission by which some people, particularly PhD students, imagine ideas to propagate over time. Learning to do something difficult entails a painful process of guided, sometimes misguided, experimentation. If I have a single guiding principle as an observer—I leave it to the reader to decide, according to their own background and inclination, whether my style of observation warrants the epithet *science*—it is this: it is essential to continually remind yourself how painful it was learning to do the thing you are now considered—by an autonomous community of practitioners? by some underspecified world-at-large?—to be expert at. It is essential to be mindful of how your outlook was molded by the periodic resentment you felt toward those responsible for guiding you, it is essential to continually put yourself anew in situations of discomfort, of not-knowing-how. It is essential to practice humility, to be skeptical that you have got it right, skeptical of your capacity to get it right.

Riki would probably tell me to lighten up.

WHAT THIS BOOK IS ABOUT

My own tastes to the contrary, a preface should say something about what the text that follows it is about.

At its center is a simple argument: the history of technology is a history of the trading off of survival strategies centered on material artifacts for those centered on *enactive* artifacts, that is, those manifest more in encultured—culturally conditioned—sensorimotor schemas than in the material residues of encultured behavior.

Since enactive artifacts do not fossilize, past instances of societies' prioritizing them over material artifacts do not come across as instances of technological innovation. To the contrary, they often come across as instances of maladaptive technology loss. This makes it difficult to imagine, say, ways out of our contemporary environmental crisis that do not entail simply substituting a new package of material artifacts, perhaps with a smaller carbon footprint, for those we have today.

Enaction is a covering term for a family of theories of cognition that start from the premise that our experience of the world is mediated not by representations "in the brain" but by our ongoing bodily exploration of the fluid boundary between self and nonself. The concept applies equally to the technological extension of our bodies in space and time. We think of technology as something that gets realized in material stuff. We incorporate this stuff into our lives and in this way our bodies change—our habits of posture and movement, patterns of muscular development and joint wear, qualities of attunement to movement, light, pressure, sound, and scent, circadian rhythms of motor vigilance, mood, and metabolic activity—the entire constellation of phenomena that in past work I have referred to as the human somatic niche. Technology serves as a scaffold for somatic niche construction.

But often, the scaffold consists not in technology in the conventional sense, that is, a repertoire of material artifacts and the behaviors associated with them, but in repertoires of enactive artifacts, shared patterns of behavior that endure by virtue of ongoing socially coordinate enaction rather than by realization in material things. Enactive artifacts are archaeologically invisible, but they are all around us. Some are ritually demarcated—working out, meditating. Others are ambient features of how we live our lives: tolerance of heat and cold, rhythms of sleep and wakefulness. Others play out over longer time horizons, such as the seasonal prescribed burning by which the earliest human inhabitants of Australia reshaped—and continue to reshape—the biota of their environment in ways that enhanced their own subsistence. Enactive phenomena are artifacts in much the same way that language is an artifact—they exist by virtue of ongoing re-creation, while their palpable traces remain largely evanescent. And, as with language, distinct recurring partial strategies for articulating body to environment spontaneously assemble into registers, and it is these registers that evolve over time.

The above, at any rate, is a fair characterization of the argument in the first half of the book, chapters 1–3. Starting with chapter 4, the argument takes something of a reflexive turn as I propose that rather than view the distinction between technological and enactive as one between opposing strategies that compete for dominance over time, we view them instead as complementary aspects of a single unfolding. This is essential if whatever insight we gain from chapters 1–3, which deal mainly with events unfolding at a remove of forty thousand to one thousand years, is to serve us in making sense of our own world. Ours is a world in which the material is so pervasive it can be difficult to see the primacy, or even the presence, of the enactive. It is when the enactive becomes methodologically invisible that we become susceptible to the crude conceptualizations of how culture mediates the human adaptation to environment that I am at such pains to counter.

Indeed, one way to read this book, as I suggested above, is as a plea for epistemic humility. This is not to say I have no conceptual—or, if you like, theoretical—ambitions. The book's title alludes to a metaphor common in the philosophy of evolution, that of the scaffold, and much of the second half consists in an effort to make sense of how scaffolds relate to another common metaphor, that of the adaptive landscape. Conceptual devices, particularly those that lend themselves to kinesthetic imagination—landscapes and scaffolds are things you can imagine yourself moving through or perhaps clambering over—serve a useful purpose. They enhance our grasp of, our cognitive access to, material that we might otherwise apprehend intellectually but without the kind of phenomenal immediacy that is essential to action. It is the difference, say, between being able to follow a mathematical proof and being able to reproduce it, or produce it anew, without feeling that you are more or less reciting it from memory. Or, to take examples that will be familiar to more readers, it is the difference between knowing *what* one needs to do to balance on a bicycle or stay afloat in water and knowing *how* to do these things. If, as many philosophers of mind hold, propositional awareness, that which allows us to formulate what we know verbally, is ultimately grounded in a more fundamental phenomenal awareness, there is often a moment of satori that accompanies the first crossing of the threshold from propositional into phenomenal. This is the moment when we first grasp something where previously we had simply been able to say it. Often this moment is transitory and we find ourselves banished again to the propositional realm, with a long slog ahead to regain that moment's effortless sense of inhabitance. But that sensation of knowing something with the body where previously we had known it in some shallower way—*with words* is a crude but useful first approximation—is real. Sometimes we experience it as a momentary lightness or as a shiver or tingling that passes through the body.

Early in life, when our propositional faculties are inchoate, we learn mainly by a sensorimotoric probing of the world and our own body—if you have ever watched a child learn to walk or talk you will have a sense what I mean. We learn, that is, enactively. Later on, much of what we learn we encounter first, sometimes of necessity, in propositional form. But this does not mean there is no role for phenomenal grasp, for enaction, in understanding these things. Often there is. Conceptual devices, deployed with a light touch, can help us bridge these two modes of apprehension.

So I find myself caught between epistemic humility and the abstracting tendency that comes with devices such as scaffolds and landscapes—not to say between those early readers who wanted this to be a book of theory-building and those who wanted it to be a series of essays in which I first narrate some episode from personal experience and then read anthropology and evolutionary theory with the reader, all the while scrupulously avoiding theoretical pretensions of my own. Is there, to mix two further metaphors, a middle way out? A deflationary approach to theory-making? This is what I've tried to model in this book. My strategy has been to offer a succession of related but nonisomorphic conceptual devices, to acknowledge that different kinesthetic metaphors may work better for different readers. To a degree, this strategy represents a knowing, even playfully self-deprecating effort to check my own weakness for theory-building. To a degree, it represents a continuation of the critical project alluded to above—for devices such as *landscape* and *scaffold*, useful in moderation, become constrictive when we inhabit them to such a degree that we lose awareness of their edges, lose awareness of the fact that they are simply devices. You know the expression about what happens when all you have is a hammer. In chapter 5, in fact, I consider the advantages, imagined and, tentatively, real, of paring back one's toolkit—toolkit in a literal, not a figurative, sense, the portmanteau of durable material things by which we ensure our passage through the day. But when it comes to our toolkit of conceptual abstractions, we could stand to be a bit more promiscuous.

BORO

There is more to be said about the rubrics, epistemological and ethical, that have guided me in the writing of this book. But I will limit myself to one further comment, this one stylistic.

In the course of writing this book, for reasons that will become clear, I started paying more attention to fashion. It was not that previously I did not care about how I dressed—I cared deeply, above all for the practical reasons alluded to above and described further in chapter 5: an epiphytic strategy is, of necessity, sartorially parsimonious. But as I wrote about the archaeology of clothing on the one hand and the role of textiles in our crisis of stuff on the other, clothing became more salient to me. No doubt living in Los Angeles also had something to do with it, as did the fact that I was spending time in an office on a regular basis for the first time in years. I started following blogs and forums, developing a taste for certain stylistic devices, modulating the ascetic functionalism that had served me well for so long. Among the tendencies—trends would be too strong—that caught my eye was *boro* (襤褸), literally *tattered*, a patchwork style descended from the strategies of repair typical of clothing and bedding in Japan prior to the advent of cheap, industrially produced cotton. Often this patchwork was applied using decorative stitching in geometric patterns, or *sashiko*, but this did little to hide the fact that patching was a strategy born of scarcity, which is to say of poverty. Recently, a handful of designers, mainly in Japan, have begun incorporating boro into clothing intended for a rather more affluent demographic. You can view this as class appropriation or as the latest in a long line of fads for simulated patina, as, say, the unfinished ("deconstructed") seams made famous by Rei Kawakubo. More favorably, you can view it as an experiment in designing clothing to degrade gracefully under use by incorporating patina directly into the look—if something never looked new, it cannot, with use, come to look inappropriately worn out.

At any rate, I have come to see this book as a work of boro. Scenes and ideas recur in patchwork fashion, versions overlapping, stitched together with elaborate embroidery, some more polished than others—now something pulled directly from a journal or a letter to a friend, now a passage of a couple hundred words that took a week's agonized effort. Fifteen years ago, when I was doing my PhD, the word *imbrication* enjoyed a moment in the interpretive social sciences. *Imbrication* is a pretentious term for overlap, as of roofing tiles—or, I suppose, quilted patches. All social phenomena had to be described as imbricated, which was simply a way of saying that they did not stand alone, that they were mutually constitutive. By describing this book as a work of boro, I have in mind something simpler: it is a bit rough around the edges, a bit unfinished, patched together, overstitched. You can read this as an admission of defeat. Or you can read it as a statement about the relationship between wholeness and cogency. Perhaps a cogent argument need not be a seamless one. Perhaps—and here I am speaking as much about the models of behavior discussed in this book as about my method of discussing them—we should be a bit suspicious of arguments that lack a certain degree of patchiness, that are too confident in their own strategy for reducing the dimensionality of encultured behavior to something analytically tractable. Dimensionality reduction—the selection of certain phenomena to focus on, the bracketing of others—is not, as too many anthropologists would hold today, intrinsically violent. But it must be undertaken with caution and with an understanding that in the long run all models are crude and provisional.

A boro writing strategy also serves as a caution against something else. One way to read this book is as an argument against *designed-object fetishism*, against the view that our problems, even when they are cultural in origin, admit of elegant technological solutions—say, in the case of consumer waste, swapping out one toolkit of stuff for a new toolkit with a smaller carbon, energy, and water footprint, with a reduced tendency to subsist in unusably degraded form in air, water, and soil, and so on. The obvious way to argue against designed-object fetishism, the reasonable way, would be to start with an account of the design profession over the past twenty years, to show how what used to be called industrial design, the design of stuff, has transformed itself into a profession whose principal object of concern is what is now called innovation strategy. There have been actors' accounts of this transformation, but, to my knowledge, no critical history. Such a work would have performed, as Riki used to say, a public service, drawing our attention to the convictions and blind spots of a community with a growing role in articulating humanity's response to its environmental predicament. Indeed, in 2014, noting design ethnographer Jan Chipchase's use of the term *popup studio* to describe what he does in the field, it occurred to me you could view the "human-centered" turn in design as the reemergence of a strategy for making knowledge about human behavior in the wild that crystallized in the 1898 Cambridge Anthropological Expedition to Torres Straits. The Cambridge Anthropological Expedition and its legacy had been a frequent theme of conversation between Riki and me and, mourning her death of a year earlier, I roughed in an essay on the expedition's significance for contemporary design. That essay represents the gametophyte to this book's sporophyte.

Alas for the reader, I have never been one to do the obvious, reasonable, or useful thing. My argument about the perils of designed-object fetishism is, like so much in this book, oblique, my proposed alternative inceptive at best. So let the boro quality of this work stand as a sketch, a *prototype*, to adopt the lexicon of design, for a design practice that did not aspire to solutions but rather to provisional if useful interventions: a patching-up, an incompleteness, something made with an awareness that everything, including our strategies for getting on in the world, disintegrates—but that perhaps we can learn to formulate strategies that disintegrate more gracefully.

Figure 1. Popup studio, 1898. Members of the Cambridge Anthropological Expedition to Torres Straits, 1898. Standing (left to right): W. H. R. Rivers, C. G. Seligman, Sidney Ray, Anthony Wilkin. Seated: A. C. Haddon. Collection of the Cambridge University Museum of Archaeology and Anthropology.

POSTSCRIPT ON THE AUSTRALIAN BUSHFIRE SEASON OF 2019–2020

It is late January as I make final revisions. In Australia, summer bushfires have consumed upwards of 18.6 million hectares (46 million acres) thus far, including 10 million hectares in the southeastern states of New South Wales and Victoria. By some estimates, more than 800 million vertebrates have died, at least thirty-four humans among them. These figures represent casualties directly attributable to fire—deaths caused by long-term excess fine aerosol pollution, habitat loss, and the stresses of displacement are more difficult to quantify. Fire season will continue another two months at least—though, to judge from recent events in California, it may no longer be appropriate to speak of fire season as if it represented a periodically bounded annual recurrence. Soon, in many parts of the world, notably the winterwet climates that are the focus of chapter 3, the whole year will be fire season.

Much of what follows concerns fire: fire as a technology of biome modification and biome maintenance and fire as a familiar, something you keep close to your body. This book does not aspire to a systematic history of prescribed burning, that is, the use of fire in landmaking, in Australia—though chapters 1 and 3 offer thoughts on the difficulties that attend efforts to write such a history. But if this patchy book has a still point at its center, something it circles around and returns to over and again, it is prescribed burning and the role it has played in Australians' strategies of inhabitance over time horizons that reach upwards of forty thousand years. Indeed, the 2019–20 bushfire season has prompted renewed attention to the practical value of Indigenous Australian habits of prescribed burning as a way of clearing the understory in order to prevent cataclysmic burns of the kind that have, over the past four weeks, destroyed large swaths of woodland in southeastern Australia. In the Northern Territory, where Indigenous strategies of land maintenance are more intact than in the southeast, Indigenous communities do take a leading role in bushfire prevention, with some individuals in more remote settlements setting preventive fires practically daily, the same way I used to tend my *myouga* in Los Angeles (actually, with far greater assurance). It is not clear whether something similar would help matters in the southeast, even if it could be put in practice.

I mention all this because it would be easy to read this book as, among other things, a paean to a lost intimacy with the unbuilt, or less-built, world—and perhaps that is fair. (As I said to one editor, in discussions about how much of my own first-person presence to keep in revision, my aim in making myself present in the text is not to emphasize my intimacy with interlocutors or with experiences germane to the argument so much as to be honest about the alienation I sometimes feel from my own body.) There is no casual intimacy in watching your home go up in flames. But perhaps the events of the past month in southeastern Australia serve as an illustration of the existential contingency that I see now I have been gesturing toward in this book—*living epiphytically, boro.* Or, to use the term I introduce in chapter 4, *foaminess.*

Kansha

In Japanese, *kansha* (感謝) means something like *appreciation* or *gratitude*. This feels more appropriate than *acknowledgment*.

At the Berggruen Institute, I give kansha for the opportunity to have spent a year working with Nils Gilman, Tobias Rees, Jenny Bourne, and Tui Shaub, my co-fellows Hélène Mialet, Hannah Landecker, and Joshua Dienstag, not to say Nicolas Berggruen, Dawn Nakagawa, Rachel Bauch, Yakov Feygin, Kristen Farlow, Alexis Dale Huang, and Nathalia Ramos, and, through the Berggruen's partnerships with USC and UCLA, Andy Lakoff and Chris Kelty. The single greatest factor in this book's coming to be was Nils's abiding conviction in the value of what I was doing. I give thanks for his support and that of the Great Transformations series coeditor Craig Calhoun.

Elizabeth Chin and Simon Penny helped Jessy and me find our way in Los Angeles. Nick Yates and Karen Merchant gave us a place to hang our epiphytes. Victor Braitberg and Ben Wurgaft made time to read and comment on drafts of this work. Among the pleasures of the time in which I wrote this book were reconnecting with Rita and Nick Tishuk and Joanna Radin and getting to know Lucy Conway and Alex Booth. At the University of California Press, I owe a debt of gratitude to Naomi Schneider, Summer Farah, and Kate Hoffman. Caroline Knapp was a singularly thoughtful copyeditor. Yoshi Sodeoka gamely took on the illustration brief for this book on short notice at the height of a pandemic.

When the pandemic caught me in New York, my parents, as so often in the past, offered me a place to shelter. I can but hope I made it easy for them.

Lynne Friedli and Marcus Salisbury went to extraordinary lengths to make Jessy and me comfortable during a high-stakes sojourn in London in September and October 2020. Their generosity humbles me, and I will be sitting with an awareness of what I owe them for a long time to come.

It was in the course of a fifteen-night stay in Onyuudani, Shiga prefecture, at the home of Harufumi and Yumie Fujimura, in September and October 2017, that I first imagined this book in more or less the form it has taken. Their guesthouse offered a place to read and take walks; equally it offered a model for a strategy of getting by that I aspire to.

Jessy: For six years you have helped me face the questions *What am I doing?* and *Where will I live?* with greater courage than I should otherwise have managed. This book is for you. May its tattered character hold the promise of further, less tattered ones to come.

1 Treadmills

For years I spent a lot of time running on treadmills. At one point I sustained a metatarsal stress fracture. When it had healed I returned to running on treadmills. The following year I reinjured my foot and had to take more time off. At length I switched from long, steady uphills with bursts of modest intensity toward the end to high-intensity interval sprinting. This had the advantage of compressing an hour's workout into ten excruciating minutes that I would dread two days in advance, and it made for less repetitive stress on the feet. But it exacerbated the fascial adhesions in the backs of the legs that come with running and left me feeling depleted. These days I run hill sprints when I can and do other things the rest of the time.

A treadmill is a machine for minimizing the biomechanical and sensorimotor complexity of locomotion. On a treadmill, there is no change in the speed of movement or the pitch of the running surface, save that introduced, at long intervals relative to the tempo of footplant (roughly 120 beats per minute for walking or 180 for running), by the runner or by the machine itself. The running surface is uniformly even in the coronal (left–right) plane. The surface does not vary in its elastic and frictive properties—asphalt versus packed earth, sand, wet moss, bark litter. Since you're not going anywhere, the distal sensory environment—what you smell, hear, and see—does not change in ways relevant to what you're doing. People may enter and exit your field of olfaction or vision, the music and other features of the acoustic environment may change, but these changes are not contingent upon your movement the way changes in the olfactory, acoustic, and visual scene are when you're running outdoors. Nor do these changes in the environment impinge on your activity, unless someone jumps on your treadmill or the roof falls in.

The biomechanical and sensorimotor simplification that the treadmill offers serves a useful purpose. In my case, it allowed me to run in times and places where I would otherwise not have been able to. But simplification has its dangers too, as witness my own history. When it comes to tolerance for different degrees of smoothness of movement, humans exhibit broad and plastic dynamic range. Too much jerkiness is frustrating, and there is something meditative about moving with a smooth, steady rhythm. But when this rhythm is mechanical, lacking in small variation, it becomes stressful.

Perhaps it is because I have a history with treadmills that I've been struck by a recent tendency, among some evolutionary biologists and anthropologists, to describe the evolution of human culture as something that unfolds—unrolls?—on a treadmill. Writing in the well-regarded *Proceedings of the Royal Society B*, anthropologists Michelle Kline and Robert Boyd note that "social learning is subject to error, and since errors will usually degrade complex adaptive traits . . . inaccurate learning creates a 'treadmill' of cultural loss, against which learners must constantly work to maintain the current level of expertise." Kline and Boyd argue that, among historically attested marine foragers (fisherfolk) of the western Pacific islands, island communities with smaller populations tended to have less complex technological toolkits—fewer tools and fewer "technological units" per tool. That is, there was a positive correlation between population size and the technological complexity of the island economy. In Kline and Boyd's analysis, this correlation is robust across three, possibly four orders of magnitude in population size—the curve does not level off for populations greater than, say, ten thousand or even one hundred thousand. What's more, in the case of the "number of tools" measure, the residuals—the divergences of individual communities from the regression curve—tend to correspond with the degree of contact ("low" or "high") that the community had with outsiders, with low-contact communities below the curve and high-contact communities above.

If borne out in a broader range of settings, this would be a significant finding. It would suggest that the key constraint on the degree to which human communities have elaborated a complex material culture over time—the degree to which they've outrun the treadmill— is not environmental (biota, climate, frequency of extreme meteorological events) but demographic. It would lend evolutionary support to recent findings that rates of innovation, as measured by number of patents issued, scale with the size of the cities where the innovation is unfolding. It would offer insight into the nature of social learning—the transmission of knowledge and skill, not to say tacit codes of behavior, from one person to another (stereotypically, parent to child) via observation, imitation, and instruction.

But what is most significant about the treadmill theory is not what it entails but how it has been formulated and defended, what it leaves out, what it takes for granted—in short, the light it casts on the constellation of tacit assumptions that underpin so much thinking, scientific and otherwise, about the nature of human niche construction, that is, the process by which we shape the environments that shape us in turn. This constellation of tacit beliefs includes assumptions about "technology," "learning," and "adaptation"—and here I've put these terms in quotation marks to show that these are what historians call actors' categories, that we must not take for granted that they mean the same thing in different contexts. The assumptions built into the treadmill theory are value-laden, which is to say that they refer, ultimately, not to things that can be arbitrated on a factual basis, such as the relative significance of population size as opposed to environmental risk in structuring foraging strategy or even the best way to measure technological complexity, but to questions such as *How should we live? What is a good life? What is flourishing?*

This is a book, then, about these questions.

The use of the term *treadmill* seems to have originated with Kline and Boyd, but their intervention comes about ten years into a debate on the role of demographic factors in cultural evolution. Most participants point to archaeologist Stephen Shennan as the instigator. In a 2001 paper, Shennan reported the results of a series of simulations designed to investigate the relationship between population size and the mean relative fitness of members of a community over time, where fitness is measured by fecundity or reproductive success (number of offspring), and reproductive success depends in turn on the degree to which an individual's scores on a panel of "craft" traits (understood to be skills essential to survival in an Upper Paleolithic community, say, weaving, butchering, or the production of stone tools by debitage or knapping) diverge from a theoretical optimum. Learning errors ("mutations") move these craft skill scores away from the optimum, though they can also, at times, move them toward it—mutation is a random walk. The question is, if fecundity, relative to a theoretical maximum, declines as an exponent of the summed divergence of an individual's skill from the optimum, how will mean fecundity evolve over time? In Shennan's simulation, over many generations, populations arrive at a stationary relative fecundity—not fixed from one generation to the next, but bounded, with a well-defined geometric mean. Relative fecundity rises with "effective population" up to an effective population of about a hundred, where it levels off. This effective population represents, in Shennan's estimate, one quarter of the census population, if you assume half the latter are kids and transmission unfolds exclusively between parents and kids of the same sex.

Shennan's model is derived from a model of trait selection first elaborated by population geneticist Ronald Fisher in 1930. It was intended as an alternative to theories of the emergence of modern human behavior that postulate a nucleotide polymorphism—a gene mutation—that gave humans enhanced cognitive capacities. But it has inspired work on a different question: *How can we account for seemingly maladaptive long-term trends in socially conveyed behavior—that is, culture?* This is where Tasmania enters the picture.

In the years since evolutionary anthropologist Joseph Henrich published his first paper on technological attrition, Tasmania has become the proverbial poster child for the treadmill model, so much so that the loss of culture under demographic isolation is sometimes referred to as the "Tasmanian effect." *Poster child* is apposite. In papers that argue for the Tasmanian effect, Tasmanians are presented in much the same way as the ostensive beneficiaries of humanitarian relief are presented to donors in the affluent world: as victims of a senseless catastrophe whose effects they lacked the wit or skill to overcome. In the case of Tasmania, the catastrophe was the flooding of what is now the Bass Strait sometime after twelve thousand years ago, which left Tasmanians cut off, so the argument goes, from the broader Australian community and unable to participate in the cultural ferment and sustained, cumulative technological innovation that comes with a larger population.

Indeed, Henrich argues, far from simply maintaining, in stasis, the technological toolkit they'd achieved prior to isolation from the mainland, Tasmanians fell off the treadmill: archaeological evidence, together with the reports of European observers from the early colonial period, indicates that by the time of European recolonization, Tasmanians had lost a number of archaeologically attested behaviors that you'd expect to find in a cool-temperate maritime environment, including the use of finfish as a source of food and the use of warm clothing. Tasmania's limited population could not sustain these behaviors in the face of the tendency of learners to make more maladaptive than adaptive errors in acquiring skills. This tendency is not a Tasmanian problem. Rather, Tasmania here is said to exemplify a general phenomenon of the social transmission of culture.

Henrich offers an elegant model to quantify the demographic threshold of maladaptive loss. The model turns on the ratio between the complexity of a given skill, operationalized in terms of how poorly people tend to do at learning it, and the degree of dispersion in learning outcomes—how many ways people get it wrong, whether maladaptive or adaptive (that is, less or more "skillful" or lower or higher in "fitness" than the individual they learned from). It is unclear what to call this second factor—perhaps "experimental fecundity" or "learning inventiveness"? Roughly, the threshold population size is exponential in the ratio between complexity and inventiveness. Put another way, the change in mean skill level over time is logarithmic in population size, with a coefficient for inventiveness and an offset for skill complexity—with enough inventive learners, you can overcome the barrier to faithful transmission posed by the complexity of the skill, though the strength of the population effect diminishes as populations grow.

It is probably clear I find Henrich's formulation brittle. In what follows I'm going to explain why. A close examination of what makes the treadmill argument problematic for Tasmania will clarify a number of concepts that are salient to contemporary discussions of our own impending climate catastrophe: technology, culture, learning, adaptation, niche. If we can achieve a more supple way of thinking about these things, perhaps we can achieve a more expansive way of thinking about strategies for responding to a pattern of environmental change that, in contrast to the Tasmanian case, is entirely of our own making.

THE GEOGRAPHY OF PLEISTOCENE TASMANIA

In the cool-arid climate of the late Pleistocene (marine isotope stages 4–2), sea levels around what is now Australia were up to

120 meters lower than they are today, and present-day New Guinea, Australia, and Tasmania formed a single land mass known as Sahul. Humans first entered Sahul sometime between sixty-five thousand and fifty thousand years ago (or 65–50 ka, *kilo annum*, to adopt a more concise notation), perhaps in a series of short hops over open water from the exposed shelf of Sunda, Pleistocene Southeast Asia. DNA evidence, autosomal and mitochondrial, favors the later date; recent archaeological evidence from a site on the present-day coast of Arnhem Land, in the north of Australia, indicates the earlier. The broader archaeological context encompasses sites from around the continent dated to between 48 and 40 ka, including one, a cave known as Devil's Lair on the southwest coast, the point on the mainland farthest from the land bridge where humans first entered the continent, dated to 48 ka. Indeed, mitochondrial haplotyping, constrained by evidence from a broad panel of archaeological sites, suggests a single, rapid colonization episode followed by rapid dispersal along the coasts, with distinct regional populations established all around the less-arid margin of the continent by 45 ka.

Humans first entered Tasmania sometime after 40 ka via a sill at the eastern edge of the Bass Plain. The late-Pleistocene Tasmanian archaeological record is exceptionally rich by Australian standards, but there remains some uncertainty about the timing of human arrival. The sill would have been exposed at some points between 50 and 30 ka, partly submerged at others. Carbon-14 dates indicate occupations at two sites, Warreen Cave, in the southwest highlands, and Parmerpar Meethaner, a rockshelter about a hundred kilometers south of what was then the northern coast, at 35 and 34 ka respectively. There has been some discussion as to whether these dates should be pushed back 4–5 ka. In some instances, comparison of carbon dates with optical luminescence dates on lithic materials suggests that the organic debris used for carbon dating has been contaminated by contact with more recent depositional contexts, yielding erroneously late dates. But there is broad consensus that signs of human occupation do not go back further than 40 ka.

Tasmania lies between 40 and 44° south. It is roughly triangular in shape, the southwest and southeast coasts meeting at a blunted tip at its southern extremity. On maps with north at the top it appears to be hanging, as if suspended from a clothesline. At present-day sea levels, the area of the main island is some 64,500 square kilometers; with the islands of the Bass Strait, including the Furneaux Group, vestige of the Bassian sill, 68,400. There is evidence for some continued occupation of the Bass Strait islands after the flooding of the Bass Plain, but the archaeological record ceases at 4.5 ka. At a later date, perhaps 2.5 ka, watercraft made of *Melaleuca* or eucalyptus bark bundled into thick cords appear on the southwest coast and the southern aspect of the southeast; thereafter, some of the strait islands were visited but not permanently reoccupied, while a handful of nearshore islands were occupied. In contrast to the Australian mainland, Tasmania's topography is rugged, with a mountainous interior with elevations above one thousand meters above sea level. These latitudes are characterized by a strong westerly air current, known as the Roaring Forties, that, in combination with the mountains, has shaped the climate of Tasmania, with the southwest receiving upwards of eighteen hundred millimeters of rain annually (at high elevations, thirty-six hundred) and the southeast substantially drier.

Evidence for the climate of Tasmania in times past comes from pollen and charcoal extracted from sediments in stratigraphic cores. The former casts light on the mix of vegetation that predominated in different places at different times and the latter on the frequency of fires. There is uncertainty surrounding these data—different kinds of plants produce different volumes of pollen, pollen from trees tends to disperse over a broader radius than that from grasses, pollen from the vegetation of bogs (wet) resembles that from moor grasses (pyrogenic—susceptible to fire). Still, in combination with evidence from marine sediment cores, pollen data show that the climate of Tasmania was cool at the time when humans first entered the island, perhaps 5 degrees Celsius cooler than at present, 6–7 degrees at peak glaciation. We tend to associate cooling episodes with aridification, but the southwest coast of Tasmania seems to have been wetter at this time than at present, at least prior to the onset of local glaciation after 25 ka.

Inland, at the Last Glacial Maximum, forest cover was sparse above one hundred meters above sea level. Even today, in many parts of the southwest interior, grassland and open sclerophyll forest (e.g., eucalyptus, *Casuarina*, and *Melaleuca*, the fire- and drought-tolerant taxa that characterize much of Australia) predominate where the climate and soil would be capable of supporting rain forest. This has led to speculation about the role humans played in keeping the land open using fire, something observed in much of Australia, including Tasmania, in the postcontact era. The degree of human intervention seems to have varied from place to place. What is clear is that grassland biomes date at least to the drying of the Last Glacial Maximum, succeeding earlier sclerophyll biomes. Grasslands were not, in the first instance, a product of human intervention. As the climate grew warmer and more humid after 14.5 ka, sclerophyll forest returned. At this stage, in some parts of the southwest but not, apparently, in the southeast, anthropogenic burning came to play a role in encouraging pyrophytic vegetation (plant taxa tolerant of fire or requiring it for habitat maintenance) and suppressing rain forest.

By luck, the caves in the southwest highlands that form the basis for our understanding of the material culture of the earliest Tasmanians are limestone, the alkalinity of which makes for extraordinary preservation. "It is not uncommon," archaeologist Richard Cosgrove writes, "to excavate 200,000 pieces of bone and 20,000 stone artifacts from less than 1 m³ of deposit." The record suggests not continuous occupation at constant or steadily increasing intensities of exploitation but rather two pulses of activity, the first between 35 and 25 ka, the second starting around 16 ka, with a lull in between (again, if you're concerned about depositional contamination, add 4 ka to the 35 ka date). This would be consistent with a diminution of activity at the peak of glaciation. At the same time, it is difficult to understand why people would gravitate to alpine uplands during a cooling episode, and in particular why they would favor caves, which are cooler and damper than open sites in the same climate. Of course there are biases of archaeological salience and preservation in play. Caves, by virtue of the shelter they provide—not to say their widely attested role as sites of shamanic activity—seem to call out for archaeological investigation, and the limestone depositional matrix has minimized the loss of organic debris to decomposition, amplifying the signal relative to other contexts. Still, there is a question here, one we will return to.

By 13 ka, all the caves of the southwest interior had been abandoned, as if, with the coming of the warmer, wetter climate that would characterize the early Holocene, the area no longer represented a productive environment. Occupation of nearby open sites does not appear in the archaeological record until late in the Holocene, perhaps less than a thousand years prior to the arrival of Europeans. Explanations for this hiatus have tended to invoke the change in biota that accompanied the change in climate: the return of sclerophyll forest, the decline in grassland. With the denser, woodier vegetation—even if the land were kept open by burning—would have come a loss of the habitat of the Bennett's wallaby (*Macropus rufogriseus*), to all appearances a key component of the human diet in this time and place. *M. rufogriseus* is relatively large as wallabies go, with females attaining sizes of 11–15.5 kilograms at maturity, males 15–27 kilograms. In contrast, say, to the eastern grey kangaroo, it is solitary and sedentary: females maintain a home range of six to eleven hectares, males fifteen to twenty, shifting the center of their range every second or third year and then no more than thirty meters. It is not difficult to understand their appeal as a food source in a cold environment that offered limited options. They make up some 70 percent of faunal assemblages from the limestone caves by specimen count. Again, there are biases in play, here of deposition and counting. There is evidence that wallabies were processed in the field, with some parts discarded, so it is possible that smaller animals, and perhaps some plant food sources, were processed and consumed in the field and never made it back to camp. Then too, specimen counts tend to overrepresent the number of individuals present. It is often impossible to say whether two bone fragments come from the same individual, and specimen counts treat them as distinct. Larger animals are further overrepresented by virtue of the fact that denser, larger-bore bones fare better under the taphonomic (tissue-decay) rigors of deposition.

Still, there is strong evidence for wallaby specialization, including the fact that in some cases wallaby long bones have been scraped of marrow. Marrow consumption represents a common strategy to compensate for a want of carbohydrate in the diet and one that finds parallels in the more diverse diet of contact-era Tasmanians.

Along with awls made from wallaby femurs, the limestone cave horizon offers the first examples of a genre of stone tool distinctive to Tasmania, known in archaeological circles as the thumbnail scraper. Thumbnail scrapers make their first appearance as early as 29 ka at one site, as late as 17 ka at another. They are oblong and bipolar (with cutting/scraping edges on either end), three to four centimeters in length, varying in size from site to site and depending on what they are made of. In the west of Tasmania, where quartz predominated, they average 2.4 grams; in the east, where they were made of chert (a sedimentary form of quartz) or silcrete (compacted soils, widely favored for blades in Paleolithic contexts), 3.4 grams. They do not change significantly in size as time goes on, but, especially in later contexts, they do sometimes exhibit delicate retouch, a precise, craftsmanlike knapping of the working edges, perhaps to sharpen a blade that had gone dull with use. This has suggested to some observers increased curation of tools and, by extension, a more mobile way of life where you held on to implements you had invested in. To other observers, the mobility-with-curation scenario feels tendentious. There is, in any case, little evidence that people sought out materials from distant sites to make scrapers or participated in trading chains for raw materials. How exactly the scrapers were used remains a bit out of focus. Use-wear studies have been inconclusive, though it is clear they were not hafted (affixed to handles or spears). Most likely they served multiple roles, among them, perhaps, the scraping of the internal faces of animal skins to make cloaks, a theme we will return to.

I have emphasized the economic lives of the early Tasmanians, and in light of our overarching project—to distinguish adaptation from flourishing, technology from culture, fitness maximization from learning—we would do well to keep in mind that not all behavior is constrained by adaptive function, at least not in the easy-to-see ways alluded to above. Three of the limestone caves investigated in the southwest highlands show no signs of occupation but do exhibit traces of a different kind of human activity: hand and arm stencils, made with red ochre, in one instance a full kilometer from the entrance. At one site, children's hands figure. These stencils date from the early Holocene. They are not accompanied by representational tableaux (for instance, animal and human figures, hunting scenes) of the types known from elsewhere in Australia and other parts of the world.

To a remarkable degree, we owe our understanding of Tasmanian social structure in the era just prior to the arrival of Europeans to the work of one individual. Rhys Jones, a Welsh-born Australian archaeologist, led the first systematic excavations in Tasmania, at Rocky Cape on the northwest coast, starting in 1963. These formed the basis for his PhD dissertation, where he proposed that Tasmanian culture had undergone a kind of simplification over time, with a loss, in particular, of finfish consumption and the use of bone points (awls) at 3.5 ka. In subsequent work, he drew a connection between the changes in the Rocky Cape economy at 3.5 ka and the isolation that had followed the flooding of the Bass Strait. Jones was not alone in arguing that isolation had led to a loss of "useful arts" in Tasmania, nor was his thesis universally accepted. His work would have a strong influence on Henrich's "maladaptive loss" thesis.

Jones was a careful reader of archival sources. It is his readings of the journals of George Augustus Robinson that provided much of the now-canonical picture of Tasmanian Aboriginal band structure and linguistic affinities in the pericontact era. The first recorded visit of Europeans to the island we know as Tasmania was that of Abel Tasman, in the course of an exploratory survey funded by the governor-general of Dutch Batavia, in 1642. Tasman named the island for his benefactor, Anthony van Diemen (it would be known as Van Diemen's Land until 1855). The first of a number of French expeditions to the island arrived in late summer (March) 1772; British presence dates to Cook's third expedition to the Pacific in 1777. Soon, French and British sealers were active in the islands of the Bass Strait. In 1803, alarmed at the increasingly permanent nature of French settlements on the island and the rapport the French seemed to have established with the Tasmanians, the governor of New South Wales dispatched an expedition to found a Crown settlement; the site chosen was the Derwent River estuary, near the southeast coast. They were soon joined by the remnants of the failed first settlement of Port Phillip (present-day Melbourne, across the Bass Strait), and the settlement was moved to the site of what would become Hobart.

By the time of Robinson's arrival, twenty years later, commercial livestock operations had flourished and settlers were engaged in a low-level frontier war with the Tasmanians, which escalated in 1826 with the declaration of the Black War. Robinson emigrated from London in 1824. In Tasmania, he established himself as a building contractor, but with the coming of the Black War he came to see himself as called to pacify the Tasmanians so as to preserve them from extermination. Commissioned in 1829 as "Aboriginal conciliator," he undertook a series of expeditions, by foot and whaleboat, to the southwest, northwest, and northeast coasts of Tasmania, which at that point had not yet been surveyed by the colonial administration. Robinson's Friendly Mission, as it was called, proceeded in parallel with a series of *cordons militaires,* collectively known as the Black Line, in which free and convict settlers, supported by garrisoned regiments of the New South Wales Corps, swept the island for bands of Tasmanians who had not yet submitted to Crown rule.

The Tasmanians were already reduced in numbers by influenza and other exotic diseases brought by sealers and settlers, not to say the loss of foraging grounds to the expanding settlement frontier. No doubt they were profoundly demoralized by the events of the past thirty years, though many showed considerable defiance. For Robinson, the best possible outcome was a reserve in the northeast of the island, where the sedentized Tasmanians would adopt agriculture and a Christian way of life. By 1831 even this was no longer feasible politically, and a reserve was established at Wybalenna, on Flinders Island in the Furneaux Group. In 1835, on the conclusion of the Friendly Mission, Robinson assumed the role of superintendent there. This position he held until 1839, when he took up the role of "Protector of Aborigines" at Port Phillip, on the Victoria side of the strait. When he departed with his family, he took fifteen Tasmanians with him. In principle this was the first step in the establishment of a more comfortable reserve on the mainland. Three and a half years later, when Robinson's successor arrived, there were just fifty-two surviving Aboriginal inhabitants at Wybalenna.

It is Robinson's journals from the first three years of the Friendly Mission, published in 1966, that afforded Jones access to archaeologically invisible aspects of Tasmanian social life. Jones inferred nine nation-like social units whose territories encompassed all parts of the island save the interior of the west and southwest. These national territories ranged in size from twenty-one hundred to eighty-five hundred square kilometers. Two were landlocked, the rest enjoyed between 110 and 550 kilometers of coastline. Their populations ranged from under two hundred to about seven hundred, with a total of some thirty-six hundred. Within each nation, landholding prerogatives resided with a number of independent bands, each composed of perhaps ten households, the whole of the population comprising some seventy-five such bands. Most seemed to have undertaken some kind of seasonal shifting of camp, in some cases of up to five hundred kilometers.

For Jones, as for practically all European observers from the immediate postcontact era, the distinctive features of a nation or "tribe" were shared language and contiguous territory. It is of course possible that glossoterritorial units of this type did exist in Tasmania at the time of contact, but from what we know of the rest of Australia, the situation is likely to have been more complex. There are further problems of interpretation, for instance, the likely mismatch between what Robinson understood by "landownership" (or the terms he translated as landownership) and the constellation of defeasible reciprocal usufruct privileges characteristic of mobile foragers with overlapping ranges. Then there is the limited nature of his own contact with many of the groups he was charged with bringing in, especially on the less-well-documented western side of the island, where his Tasmanian guides from the colonized side were often unknown or unwelcome. Even if we take his record at face value, we must still consider the more than two generations between 1772 and 1829, which undoubtedly witnessed a restructuring, not to say destructuring, of Tasmanian social life all but invisible to even the most sensitive and empathic French and British observers. It did not take long, for instance, for European sealers to establish intimate domestic ties with Tasmanian women on the north coast, where seal hunting was a women's industry. The nature of these relationships—coerced, free, transactional—and their effects on Tasmanians of both sexes have been debated. What is incontrovertible is that the setting for Robinson's observations—and for Jones's inferences—was not a stable social and economic milieu but a population under siege.

It is a testament to the dearth of ethnographic records that, despite this, and despite subsequent events that would land him with accusations of willfully denying the continuance of the Aboriginal Tasmanian community down to the present, Jones's version has become the standard account of late-Holocene social structure in Tasmania. For late-Holocene economy we have other sources, notably the heroic efforts of Betty Hiatt, whose bachelor's thesis, "The Food Quest and the Economy of the Tasmanian Aborigines," surveyed a wide array of published and unpublished firsthand observations of Tasmanian economic life going back to the 1770s.

The picture that emerges from Hiatt's survey is marked in its contrast to that afforded by the limestone-cave record for Pleistocene Tasmania—which, recall, suggested specialization in one species of large wallaby. The diet was diverse, drawing on a wide range of flora and fauna. Plant-source foods, Hiatt estimates, played less of a role than in the diet of mainland Australians (then estimated at 70 to 80 percent of the diet by energy; more recent estimates suggest even greater reliance on plants in certain times and places). This could be down to observer bias, though her sources make frequent mention of roots, underground storage organs, and a type of fungus whose growth seems to be stimulated by fire. Key animal-source components of the diet include wallaby, possum, seal, puffin, duck, and emu, along with the eggs of the last three. Smaller birds and marsupials the Tasmanians trapped with snares; seals they hunted with clubs. Larger macropods such as kangaroo and carnivorous marsupials such as the thylacine were uncommon in the diet but not unheard-of.

Notably absent from the ethnographic record were observations of the use of finfish. By contrast, marine mollusks such as abalone had a central role in the diet, appearing almost as frequently in Hiatt's sources, by number of mentions, as plant-source foods and macropods; on the coasts, mollusks are more common than either plant sources or macropods. Early observers describe women and girls diving for abalone and crustaceans; Labillardière, the naturalist with the D'Entrecasteaux expedition, which visited Tasmania in 1792 and 1793, notes their use of spatulas to scrape gastropods from the undersea rocks.

Hiatt summarizes her findings by contrasting the foraging strategies of women and men: "Reliable items in the diet of the Tasmanians were supplied by the women; these included shell fish, crustacea and vegetables. Women also caught possums, seals and collected water and wood. They did most of the cooking and built the houses. Tasmanian men were responsible for the less reliable elements in the diet. They captured animals that had to be stalked, for example, kangaroos, wallabies, wombats and seals." This concords with observations in other immediate-return (that is, little storage of food for delayed consumption) foraging economies, notably, despite differences in climate and biota, the deserts of Western Australia. Women tend to adopt strategies that minimize the risk of coming home empty-handed, and in doing so they end up providing the bulk of the diet. Men, by contrast, gravitate to strategies that maximize the potential for windfalls (high yield for a single outing) and that afford opportunities for displays of ballistic prowess and physical courage.

This is not to say women were lacking in physical courage, as witness a seal-hunting technique observed in 1816: six women swam out to a cluster of offshore rocks where a pod of seals were dozing, lay down among them, scratched themselves in seal-like fashion, and lay still for close to an hour, the sea washing over them, until the seals, apparently, no longer found them salient. Then they jumped up and clubbed the seals on the nose, killing two each, and swam back to shore with them, making two trips.

Nor should the contrast between women's and men's strategies be taken to show that men were more interested in competitive display than in feeding the coresidential group. The reasons for the difference are complex and debated. Among other things, there are times when a windfall strategy makes sense—say, when the extrinsic risk of a shortfall (that is, the component of risk due to disadvantageous environmental circumstances, not to choice of foraging strategy) is exceptionally high, so that even a risk-averse strategy is no longer a secure bet. For these times, you'd better have maintained the skills and toolkit needed to maximize potential windfalls or everyone could starve. On one level, the men's strategy represents a form of costly signaling—by expending greater effort than necessary you demonstrate your commitment to the group. On another level, the men's strategy represents a skills reservoir for periods of environmental stress. If nothing else, the persistence of high-risk/high-yield strategies in the face of their comparatively small contribution, at most times, to the community's food energy budget should make us cautious about describing foragers' behavior as an exercise in maximizing fitness, at least fitness as it is conventionally understood.

In addition to clubs (of native design as well as those influenced by the clubs of European sealers), snares and bird-traps, woven bags and baskets, water bladders made of kelp, the spatulas women used to collect marine gastropods and barnacles, and bark-bundle canoes, Tasmanians made use of one-piece spears in the hunting of large, mobile marsupials and perhaps emu. Not observed in their toolkit were hafted spears and axes, spear throwers, or boomerangs—mainstays of the hunting toolkit on the mainland—nor were these things observed in the archaeological record. Should we expect to find them? The topography and biome structure of Tasmania is rather different from that of the mainland—vertically oriented, with more closed-canopy forest, and an absence of the open plains where projectile weapons make sense. A more interesting absence, noted by many early European observers, is clothing.

Of all the things "missing" from the Tasmanian technological repertoire, perhaps none has so unnerved outside observers as the fact that, by and large, Tasmanians did not wear clothing. This is not, it must be emphasized, because early European observers were scandalized by nakedness. If anything, they expected it of the people they encountered on the far side of the world. Nothing would be more disconcerting than to arrive in a new continent, secure in the belief that you are emissaries of a more advanced race, to find people who exhibit a similar standard of decorum and sophistication with

Figure 2. Bark-bundle canoes, eastern shore of Schouten Island, 1802. Color engraving by Charles Alexandre Lesueur, Baudin Expedition. Originally published in François Péron, *Voyage de découvertes aux terres australes, partie historique* (Paris: Imprimerie impériale, 1807–1816), plate 14. The caption to the original plate reads, "N.B. The spit of land on which the large canoe rests does not exist in reality; the sea at this location is completely open." Reproduction courtesy the National Library of Australia, NF 919.404 L644.

regard to bodily comportment to your own, the same sense of shame at public nakedness, the same elaboration of decoration, style, and the use of clothing to signal distinctions of class and social role. From the point of view of the colonizer, absence or crudeness of clothing on the part of one's opposite number was reassuring, for it signaled absence of political institutions of a sort one would be obliged to recognize as equivalent to one's own.

What made the absence of clothing among the Tasmanians disturbing was that Tasmania was cold, not to say damp. It did not seem possible that people could live comfortably in this climate without clothing.

Like so much of the evidence we've encountered thus far, early reports of "nakedness" among Tasmanians present problems of interpretation. First is simply what European observers meant by "naked"—complete absence of any kind of body covering, or simply a failure to cover socially proscribed parts of the body, so that someone wearing a cloak, say, but with the genitals and, in the case of women, breasts exposed, would rate, for some observers, as "naked"? In some cases it was not clear what should count as clothing as opposed to portable storage, as with the wallaby capes worn by women more, apparently, to carry children and gear than for protection from the elements. Then there are biases of observation. In Tasmania, in the pericontact era, most observations come from summer and autumn, when sailing conditions were best, and most of those whom the visitors encountered were men (sampling is more even for the period from 1803, when British observers were present in the southeast year-round and had contact with Tasmanians in a wider range of situations than those typical of ritualized trade encounters). By contrast, most early observations from the northern part of the Australian mainland, where the climate is monsoon tropical and Europeans did not establish a year-round presence until much later, come from the cool-dry part of the year.

Taking these potential confounds into account and systematically surveying early observers' reports for remarks on the presence or absence of clothing, the archaeologist Ian Gilligan has concluded that, on the Australian mainland, there is a clear thermal trend in the observed use of clothing: as latitude increases, so does the tendency among observers to report the presence of clothing, while reports of nakedness decline. In Tasmania, however, the trend reverses: fewer reports of clothing, more of its absence. Was this stasis, regression, or adaptation by other-than-technological means? Let's consider the archaeological evidence.

The archaeology of clothing presents unique difficulties. Clothing, whether of skins or textiles, does not fossilize well. Sometimes you can infer its presence from the presence of beads (small artifacts of bone, shell, or wood, often with incised designs) that have been pierced or drilled in ways suggesting they were used as toggles, buttons, or decoration. In other cases the presence of tools, lithic or bone, for working hides—scrapers, handheld blades, awls, eyed needles—offers a proxy for clothing industries, with eyed needles in particular suggesting the use of complex, fitted clothing made of multiple pieces of skin sewn together with gut or nerve. Use-wear studies—close observation of patterns of abrasion on the working surfaces of recovered artifacts, in combination with experimental studies to recreate the wear characteristic of different forms of use—can enhance inferences from tool assemblages, though as we saw in the case of Pleistocene Tasmanian thumbnail scrapers, multiuse tools tend to yield ambiguous use-wear evidence. Sometimes faunal assemblages show distinct signs that the community in question targeted fur-bearing animals. Occasionally (though not in Tasmania), you find parietal or plastic depictions (petroglyphs, figurines) that show people wearing clothes. Even in the best of circumstances, the evidence remains indirect even by archaeological standards. In the case of figurative art, there is the added confound that the presence of clothing could signify social status, not widespread use, something you might expect if the persons depicted were gods, priests, or chieftains.

This last point suggests a more basic theoretical problem: Should we expect to see clothing arise first in response to the thermal demands of the environment or in response to an elaboration of social and ritual life? The thermal hypothesis may seem obvious until you note that climates that we consider intolerable in the absence of clothing may, with acclimatization, be comfortable. (Rather, habitual use of clothing, especially fitted clothing that offers an effective barrier to wind, promotes deacclimatization to cold.) Experimental and observational data suggest that the threshold of safe exposure for humans is around –1 degrees Celsius; with acclimatization and adaptation (discussed below) regular exposure to temperatures of –5 Celsius might be possible. You could imagine scenarios in which clothing evolved to meet some other role and was then exapted to facilitate migration into colder climates.

In Tasmania, the evidence favors a thermal hypothesis, both for the emergence of clothing in the Pleistocene and for its disappearance in the Holocene. For the Pleistocene, indirect evidence includes the thumbnail-scraper-and-bone-point industry of the southwest limestone caves along with the specialization in large-bodied wallabies observed in the same assemblages. Use-wear analysis for the scrapers suggests multiple uses, among them the scraping of skins necessary for the making of capes and cloaks. That for the bone points indicates they were used, among other things, to pierce animal skins that had already been dried, again consistent with a cape/cloak industry. No evidence has been found of the manufacture of complex (pieced, fitted) clothing. Here is where we need to expand our panel of climate proxies. Temperature and humidity proxies tell us something about climate in times past, but in the case of thermal stress often the determining factor is windchill.

In Tasmania at the Last Glacial Maximum, recall, temperatures were 6–7 degrees cooler, on average, than at present. Apart from the southwest coast, the island was drier then too, with annual precipitation perhaps as low as half what it is at present. Wind speeds, estimated by the mass curves for pollen and dust particles trapped in paleosols (fossil soils), may have been eight to ten kilometers per hour stronger than today. Even if winds were no stronger than at present, there was, as we've seen, less woody vegetative cover, leaving humans and other large-bodied endotherms more exposed to the effects of windchill, even in a drier climate—and, in the case of humans, with less building material for shelter and less fuel for fire.

These climate proxies, especially those for wind speed, are coarse-grained. But taken together, they suggest a correlation between windchill and occupation of the limestone caves of the southwest interior. The correlation is not exactly linear—it incorporates a lag, both between cooling in the run-up to the Last Glacial Maximum and the occupation of cave sites and between warming and the abandonment of cave sites. Some of these caves, recall, appear to have been abandoned during the coldest part of the glaciation, or at least to have had fewer occupants then. But it appears that sheltered sites were sought out for the thermal security they offered, even though they were at higher elevation and in a colder, damper part of the island. Here is the answer to the question we were left with earlier about what attracted people to the caves of the southwest interior at the coldest period in the human occupation of Tasmania. Indeed, of the caves that were occupied, none faced south or southwest, the direction of prevailing winds.

In Tasmania, windchill never exceeded the threshold beyond which fitted clothing would be required for survival. A combination of site selection and simple (cape/cloak) body cover afforded sufficient protection from the elements while at the same time supporting adaptation to the cold, so that, with the climate amelioration that followed, the return of forests, and migration to the coasts, clothing no longer served the urgent purpose it had before. Around 3.5 ka, recall, bone points disappear from the record altogether. What did not disappear was the exceptional cold tolerance attributed to the Tasmanians by early European observers.

Physiological adaptation to cold in humans remains incompletely understood. It is clearly both developmental (epigenetic) and transgenerational (genetic and epigenetic) in nature. In terms of mechanism, it takes at least two forms. One is the retention of thermogenic brown adipose tissue into adulthood—or, perhaps, the expansion of brown adipose tissue under recurring cold exposure. Brown adipose reserves may provide a 10 to 15 percent boost to the metabolic rate under cold stress. A perhaps more significant component is an elevation of the basal metabolic rate, mediated in part by thyroid function. (The precise mechanisms remain unknown, though there is speculation of a partial decoupling of respiration from oxidative phosphorylation, so that energy is released as heat rather than stored as ATP. This would be consistent with a role for brown adipose tissue, which is distinguished by mitochondria that produce UCP1, a protein that facilitates respiratory decoupling.) Of course, elevated metabolic rate does not come free, especially in environments where sources of food, carbohydrates in particular, are limited. Possible tradeoffs for elevated basal metabolism include lower total fertility among women and slower body growth during development. This last point suggests that morphological traits associated with communities with a long history in cold climates—a more compact, endomorphic build with relatively short limbs—are at least partly developmental in origin. These traits reduce the surface area of the body exposed to cold and wind, so over the course of many generations in a cold environment they may be selected for genetically too.

But the most important adaptations are neither physiological nor morphological but behavioral. These comprise three dimensions, all of which are difficult to recover at a historical remove, though first-hand observer accounts help. The first is semiotic awareness: knowing how to read the weather, the behavior of other animals, and interoceptive signals from one's own body. The second is environment modification: knowing how to build windbreaks and keep fire handy, both domains where Tasmanians greatly impressed their uninvited guests. The third is the most difficult to see clearly at a distance of two hundred years, let alone 2–10 ka, but it may be the most important. It is the emotional dimension of adaptation to the environment, in this case nonaversion—being at home in one's skin in the cold. This can be learned, preferably early in life, and it can be modeled—children take their cues about how to feel about a challenging situation from nearby adults, especially regular caregivers. Within a community, no doubt there will be some variation, perhaps normally distributed, in how good people are at not letting the cold get to them. But it's not clear that it is meaningful to speak of this kind of skill as having a measurable, or even rankable (that is, ordinal-measurable) complexity in the manner envisaged by proponents of the treadmill theory.

Just above I said the Tasmanians impressed early European observers with their use of fire. Earlier I canvassed evidence that some of the wildfire implicated in Tasmania's unexpectedly dry biome structure was anthropogenic—of human origin. The ethnographic record of Tasmanians' use of fire for land modification dates at least to the 1770s, possibly to Tasman's visit in 1642. Some accounts include descriptions of flint-like stones that the Tasmanians kept on hand for starting fire by percussion, and others include mention of friction methods similar to those found on the mainland. But there is substantially more evidence for the careful curation of fire—via smoldering pieces of bark kept on hand when they were traveling and clay hearths kept in the bark-bundle canoes. This has given rise to speculation that, in fact, Tasmanians could not make fire at will but were constrained to curate fire got by other means, say by preserving it from a bushfire caused by lightning or by borrowing it from a neighboring group. Robinson's journals are frequently cited in support of this hypothesis.

After reviewing the limited evidence and expressing exasperation that so many commentators have thought it "reasonable to assume that the Tasmanians preserved fire for 12,000 years [i.e., since isolation, not since first occupation] depending entirely upon lightning strikes as the means to renew it," anthropologist Beth Gott briskly summarizes what we can know: "Fire-making was difficult in the damp Tasmanian climate, and the preference was to carry fire from place to place, but the Tasmanians did know how to make fire. Fire-making may have been a skill possessed only by certain members of the group. Making fire by percussion seems to have been limited to an area of southern Tasmania around Bruny Island. The fire-plough was the most likely other method of fire-making, and the drill method is the one most likely to have been acquired from mainland sources [i.e., postcontact]."

To this I would add a couple of points. The first is this: even in dry climates, starting a fire by friction is a lot harder than it looks and not something done casually. Damp climate may have pushed Tasmanians toward a fire strategy that emphasized skillful curation over frequent ignition at will simply because, in that climate, curation was more reliable and less effortful. To most readers this will seem alien. Curation seems fraught with the constant peril of setting oneself and one's living space on fire. But accidents of this sort are not noted in the ethnographic record at any point in the sixty years from the onset of regular contact to removal to Flinders Island. So if the subtext of insinuations that Tasmanians could not make fire at will is that they were not behaviorally modern, because behavioral modernity entails the control of fire, then we need to rethink what we mean by control. Certainly Tasmanians' comfort with curated fire suggests they were better habituated to it and had more supple control over it than we.

A second, related point is that, historically, gaining control over fire was a process, not an event. At time depths on the order of 400–800 ka, it is difficult to separate the elements of this process in the archaeological record—habituation to wildfire, opportunistic use, curation, ignition at will—and we tend to imagine a one-time "discovery" of ignition at will as the moment when humans mastered fire. But ignition at will comes at the end of a long period of habituation and use. From a behavioral and cognitive standpoint, it is not the most important part of the human relationship with fire. It is simply the part that yields unequivocal fossil evidence. As we've seen, archaeological visibility is not a reliable proxy for behavioral complexity.

BORO: A SENSE OF SCALE

Let's revisit that last paragraph. *—a process, not an event. At time depths on the order of 400–800 ka . . . we tend to imagine a one-time "discovery" of ignition at will—* Here we see a narrative phenomenon that will recur in what follows: a shift in scale, from the macroscopic to the microscopic, as here, or, sometimes, from the microscopic to the macroscopic. We've encountered this a bunch of times already:

Perhaps it is because I have a history with treadmills that I've been struck by a recent tendency . . . to describe the evolution of human culture—

a pattern of environmental change that, in contrast to the Tasmanian case, is entirely of our own making—

Physiological adaptation to cold [is] both developmental (epigenetic) and transgenerational (genetic and epigenetic) in nature—

and we'll encounter it again in chapters 2, 3, and 4. I won't flag it every time, though I will point it out here and there. Sometimes the shift will be in the scale of time of the phenomenon under consideration, sometimes in the scale of space, sometimes in the scale of social complexity—essentially, the size and topological complexity (degree of connectedness, uniformity of "neighbor" or nearness relations, etc.) of the social network manifest in the population whose behavior we're interested in. This last kind of scale, that of social complexity, will prove the most vexing. At the end of chapter 4, when we've grappled with the challenges of inductive generalization across scales of social complexity, I'll look at how one might (that is, how biologists, anthropologists, and philosophers have) reasoned about the time course of social complexity—how strategies for scaffolding the expansion and thickening of social networks take hold in a community, and with what implications for how that community fashions and refashions its niche.

Scale presents a challenge that is both narrative and epistemological, that is, a challenge for our theory of knowledge. The narrative challenge, as we've seen, is that shifts in scale can unfold in the space of a single sentence, and while it would be unwieldy to mark these shifts every time, it takes practice to be aware of them. This is as true for the writer as it is for the reader—even after writing two books that traffic in the deep-historical contextualization of contemporary behavior, I rely on others to point out instances where I've elided a jump in scale that is, perhaps, not warranted by the evidence. This gets to the second challenge posed by scale, the challenge for our theory of knowledge. *How do we know when generalizations across scales are warranted?* Ideally, you'd like to have a panel of cases, varying only in scale—one kind of scale—and in fine-grained steps at that, that you could compare to see if claims argued at one end, generally the "simpler" or smaller end, hold all the way up. But history does not unfold in a series of reproducible experiments. Inevitably, reasoning with scale entails judgment, interpretive discretion, the sort of thing born of long immersion in a body of evidence in dialogue with others whose perspective—by dint of accidents of personal history or disposition—differs from one's own. Does this mean judgments about history that entail inductive generalization across scales are unreliable or "unscientific"? No. It simply means that the process of inductive generalization, be it from a single archaeological debris assemblage to the behavior of a society or from the behavior of a foraging society to that of an urban society, is imbued with the values of those doing the generalizing. This is something we've seen above. We will see it again in the chapter that follows, when we look at how proponents of trait-transmission theories of cultural evolution formulate their models.

An indulgence: in my struggle to get to grips with the conjoined narrative and epistemological challenges of scale, I have been helped, not for the first time, by literature. Consider the following, the opening passage from Gerald Murnane's 1974 novel *Tamarisk Row*:

> On one of the last days of December 1947 a nine-year-old boy named Clement Killeaton and his father, Augustine, look up for the first time at a calendar published by St Columban's Missionary Society. The first page of the calendar is headed *January 1948* and has a picture of Jesus and his parents resting on their journey from Palestine to Egypt. Below the picture, the page is divided by thick black lines into thirty-one yellow squares. Each of the squares is a day all over the plains of northern Victoria and over the city of Bassett where Clement and his parents set out and return home across the orange quartzy gravel of footpaths and the black strips of bitumen in the centres of streets, only seldom remembering that high over a landscape of bright patterns of days the boy-hero of their religion looks out across journeys of

people the size of fly-specks across paper the colour of sunlight in years he can never forget.

What makes this passage, not to say *Tamarisk Row*, so extraordinary is Murnane's seemingly effortless shift in scale, from the calendric representation of days to the days themselves as they unfold "all over the plains of northern Victoria," with features in the representation—the grid, the illustration of the Flight into Egypt—finding counterparts in the world at large—the asphalt of the streets, the god whom the inhabitants of Bassett imagine looks down on them as they go about their days. So much of *Tamarisk Row* consists in Clement Killeaton's recreations of a landscape—the interior of Australia, not to say the social landscape he observes at school and in his father's interactions with horse-racing bookmakers—in the dusty backyard of his home. There is a kind of cosmological analogism at work here similar to that often attributed to classical China, where the body and conduct of emperor and state were understood to enact, in microcosm, the substance and unfolding of the cosmos—indeed, where the comportment of the emperor and the conduct of affairs of state were understood to participate in the governance of the cosmos. My own use of shifts of scale is more modest in aim: it seeks to allow events in the deep past to serve as design fictions for possible futures, our own—the question of who the *our own* refers to is something we'll come to in the second half of the book—among them. In this, I have tried to be cautious.

2 Scaffolds

Early in the previous chapter I promised to explain why I find demographic models of cultural complexity brittle. First I should explain what I mean by *brittle*. By *brittle* I mean *founded on assumptions that grossly underestimate the range of mechanisms at work in the evolution of behavior*—in other words, liable to break if it turns out those assumptions are not met. I've suggested that what is most interesting about these models is not the predictions they make but the ways in which they are brittle, for this brittleness echoes widespread biases in how we think about the relationships between technology and adaptation and between adaptation and environmental change when we consider our own world and its fate. To put it crudely, these biases favor a stuff-centered view of adaptation—adaptation is something mediated by technology—and a view of environmental change as something extrinsic to behavior.

We've made a start. In the last chapter I canvassed some of the ways in which the treadmill model, in particular in its application to Tasmania, is empirically brittle in its reading of the archaeological and historical record on matters such as clothing and fire. Toward the end of the last chapter I edged into some of the conceptual problems of the treadmill model, specifically its one-dimensional view of behavioral complexity. Recall, in the versions of the treadmill model we've seen, behavioral complexity is understood as having some underlying meaning indicated by the number of parts in a tool or the number of steps in a process for making a tool, preparing a meal, et cetera. In the absence of a uniform way to measure this kind of complexity, the models operationalize it in terms of the degree to which individuals get something wrong when they learn the skill. A quick consideration of domains of everyday behavior such as thermoregulation and the use of fire suggests how limited this way of thinking about behavioral complexity is. In the case of thermoregulation it seems doubtful that variation from one person to the next in how well an individual "acquires" emotional tolerance of cold by observing adults is really a measure of its complexity—or, in fact, that complexity is a meaningful rubric for characterizing the effort entailed, over the course of childhood and adolescence, in developing cold tolerance.

In the case of fire, I suggested that there's a tradeoff between behavioral skillfulness and technological complexity. A community that prioritizes the skillful curation of fire over ignition at will retains a much more supple behavioral repertoire for interacting with fire, and it does so with a minimally elaborated material apparatus—in Tasmania, strips of bark and lumps of clay. In this case, technological elaboration in the form of new instruments of ignition would likely go hand in hand with a regression in behavioral skillfulness, as the skillful curation of fire diminished in value. (In passing, note how the *would likely go* entails a shift in scale, from isolated to recurring episodes of innovation and from episodes of innovation to a *regression in behavioral skillfulness* that would play out over a generation or longer.)

We saw something similar in the case of the gendered quality of foraging strategies. There, it appeared, at first glance, that men were

doing something the hard way simply to show off. That is, they clove
to a strategy that demanded great skill and, by implication, a great
investment of time and effort in learning to be skillful, not to mention the risks of injury associated with hunting large, mobile prey,
even though an alternative was available that yielded more consistent returns from one outing to the next at lower risk to life and
limb. (Note that I am not suggesting the men's strategy demanded
greater skillfulness than the women's, simply that the ratio of
required skillfulness to expected returns was higher in the men's
strategy than in the women's.) One thing I did not emphasize in the
earlier discussion is that immediate-return foraging societies tend to
be highly egalitarian. You are expected to share whatever you've collected or killed not just with immediate kin but with everyone in the
camp or coresidential group. In this context, showing off—"costly
signaling"—may be strategic in itself. Costly signaling serves as a way
of advertising your commitment to the group, since, whenever you
do succeed with a high-risk/high-yield strategy, you will not get to
keep much, if anything, of what you've caught. But I suggested that
there's more to doing things the hard way than showing off. Adopting a strategy that prioritizes skillfulness relative to expected benefit
may also serve to maintain a reservoir of skills that appear superflu-
ous most of the time but become economically essential in times of
environmental stress.

Critically, the behavioral repertoires that become invisible when
you define behavioral complexity strictly in terms of things that are
easy to see in the archaeological and historical record are bodily repertoires. In this chapter, as we turn to the conceptual problems of
current models of change, demographic and otherwise, in cultural
complexity over time, bodies will move to the center of the discussion. In principle, they were there all along. All the activities enumerated in the treadmill model are things that demand bodily effort,
whether it's a child observing an adult at work and imitating her
behavior or the same individual, now an adult, diving for abalone,
clubbing a seal, scraping a wallaby skin, assembling a windbreak, or

building a fire from embers carried from the last camp. But somewhere along the referential chain linking firsthand experience to firsthand or archaeological observation to reading about that observation to formulating a model of skill complexity—*note the ladder of scale running from event to representation*—the body drops out, and with it, a clear understanding of just how high-dimensional behavioral complexity is.

Let me be clear, I am not arguing that modeling is intellectually vacuous because models can never match the richness of experience. Modeling is an important tool for reasoning about our world and identifying its principal dimensions, the things that matter most in determining differences, from one case to another, in how things turn out. But models cannot be so parsimonious that they leave out most of what is important about the phenomena in question, such as the fact that human behavior is both supported and constrained by the morphological and physiological properties that characterize human bodies.

Bodies, that is, serve as scaffolds for behavior—they both support and constrain its growth and development. Later in the chapter I'll start to lay out a competing vision for how to think about change over time in a community's strategy for getting a living and responding to environmental change. This competing vision will demand an alternative matrix of concepts to that implicated in the treadmill model—fitness, skills transmission, adaptation. Scaffolds and scaffolding will play a key role in my proposed alternative.

There's something else that drops out in treadmill models, and that is the environment. We've seen this in the way a brief consideration of changes in environmental circumstances ("extrinsic risk"), above all changes in climate and biota, calls into question the strength of the population effect in constraining a community's capacity to maintain an economic repertoire of a given complexity. But there's another sense in which the environment drops out of demographic models, and that is that they ignore the degree to which communities create the environment that then serves as the stage for their activity, economic and otherwise. This phenomenon, in which selective cues run from the community to its environment as well as from environment to community, is known as *niche construction*. We've seen it in action in the ethnographic and palynological (pollen and charcoal) evidence that Tasmanians used fire to keep the land open. We'll have more to say about the use of fire as an instrument of biome modification later on. But niche construction can also refer to how a community shapes the panel of resources it relies on within its environment, so for instance we can speak of a nutritional niche. Case in point, the absence of finfish in the Tasmanian diet at the time of recolonization.

Recall that Rhys Jones noted an inexplicable stratigraphic horizon in the debris field at Rocky Cape on the northwest coast of Tasmania. At 3.5 ka, two things that had been common in earlier assemblages drop out: bone points and fish. Bone points we considered in the last chapter. But the loss of fish, in Jones's view, was the real marker of an isolation-induced dimming of the cultural spark. Other excavations have since confirmed a sharp diminution in the incidence of fish scales in middens from about 4 ka. The early ethnographic evidence is a bit difficult to interpret. But it suggests that, prior to removal to Flinders Island, finfish played a limited role in the diet, in many places none at all.

Should this give us pause? Recall that in the Pleistocene limestone cave occupations, faunal assemblages feature wallabies and marsupials whose long bones have been broken open and the marrow scraped out, suggesting a strategy of using marrow to make up for a dearth of carbohydrates in the diet. There is some evidence that even in the warmer climate of the Holocene, carbohydrates were lacking in the Tasmanian diet. In these circumstances, gastropods such as abalone represent a valuable source of carbohydrates, while the lean finfish to be had in Tasmania would only exacerbate the problem of excess protein in the diet. If in fact Tasmanians shifted away from reliance on finfish after 4 ka, this may represent, in part, a response to an increased difficulty getting plant-source carbohydrates under the cooling pulse of the later Holocene.

So far we've seen three ways in which treadmill hypotheses fail to capture the phenomena they aspire to model. One is that they conflate complexity and skillfulness. They assume that more-complex behavior is more demanding of skill. They further assume that the key component in the social reproduction of skillful behavior is observational learning, and that more-skillful behavior is more difficult to learn, say, because it entails a broader array of concurrent subtle gestures that learners, above all children and adolescents, will have difficulty picking up on and reproducing in their own bodies when they observe skilled models. If this sounds nebulous—if you're having trouble forming an image in your mind of the scene of social learning, the moment when a young observer forms inferences, correct or otherwise, about the sequence and dynamics of the cascade of motor impulses by which a skilled model throws a spear, applies percussive force to a quartz flake to retouch its edge, or builds a fire from embers—you're not alone. Treadmill models and, in fact, to date just about all evolutionary models of the relationship between social learning and cultural complexity, are maddeningly vague about the bodily and dyadic (two bodies in coordination) dynamics of social learning. This is by design, since the purpose of these models is to connect events on one time horizon, the episodic horizon of observational learning, to events on a much longer horizon, that of the acquisition and loss of complex behavior over tens or hundreds of generations—*that is, the purpose of these models is to enact a shift in scale.* The problem is not simply that these models omit fine detail at the episodic end of the continuum. The real problem is that the assumptions these models make about what happens at the episodic end—for instance, about the relationship between observing skillful behavior and reproducing it in one's own body—are crude, quasideductive, and unwarranted.

Combining these two assumptions—complexity equals skillful-ness, correct inference from observation is the key step in the social reproduction of skill—we get a third assumption: more complex behavior faces higher barriers to faithful transmission from one generation to the next.

There is more to take issue with before we abandon treadmills, but let's pause here. We've seen enough to understand both why treadmill models conflate complexity with skillfulness and why this is problematic. The first *why* is that treadmill models, and the broader class of evolutionary models of cultural complexity of which they form one genre, grow out of investigations into technological complexity that take the complexity of tools themselves as a proxy for the complexity of a tool-using culture. The complexity of tools, in turn, tends to be measured either by the number of components subsumed in the tool in its working state or, in more sophisticated formulations, by the number of junctures by which those compo-nents are configured into the whole. So, for instance, in "count the component" models, a thumbnail scraper would rate a complexity of one (one component) while a hafted spear would rate a two (two components, the shaft and the point, bound together by some com-bination of adhesive resins and mechanical bindings).

You could take issue with these ways of ranking tool complexity (for instance: Why should a one-component tool that demands skill-ful curation over time rank as less complex than a multicomponent tool whose components are easy to make and frequently cast aside? Is a snare less complex than a hafted spear by virtue of its having just one component, when the way that component is joined to itself, and the fact that the shape of the tool varies in the course of its use, sug-gests a richer form of causal reasoning than that implicated in the spear?), but there's a bigger question at stake: How do you translate tool complexity models of this type to skill complexity? This is some-thing treadmill proponents have never made clear. A clue comes from Shennan's use of the phrase "craft skills," which conjures up scenes of tool manufacture, where the number of steps in the manufacturing process might take the place of the number of units in the finished tool. In fact, there's a rich tradition in cognitive archaeology, going back to the work of the French archaeologist André Leroi-Gourhan in the 1950s and 1960s, of "operatory chain" *(chaîne opératoire)* analysis, which seeks to bring precision to the cognitive complexity of different kinds of behavior. Operatory chains provide a device for getting to grips with the number of things an individual, dyad, or group needs to keep track of simultaneously at each stage in some process, be it manufacturing bark-bundle canoes, erecting a windbreak, or hunting wallaby, but these chains cannot be reduced to an ordinal ranking ("This process is of complexity 1, that one is of complexity 2").

In any event, cultural complexity models never refer to the operatory chain literature, and they make no implicit use of its findings. Instead, they simply assume, tacitly, that the complexity of behavior can be described in a manner analogous to that of tools. What's more, they take tools themselves as paradigm instances of the media of transmission of behavior from one generation to the next. That is, they take technological artifacts in the everyday sense—things you can hold in your hand—as the principal scaffolds of complex behavior.

We've seen that skillfulness and technological elaboration often stand in tension, with the one suffering as the other flourishes—as with the curation of fire. So, from a historical standpoint, conflating skillfulness and tool complexity makes for poor models. But there's an even more basic problem, which is that the complexity of tools, or even of toolkits (assemblages of tools used in concert as part of an overall economic strategy), is a poor proxy for the richness of the tool-using culture in which those tools are embedded. The archaeologist Miriam Noël Haidle offers an elegant image: "As the organism of a fungus cannot be understood by studying mushrooms which appear at the surface for limited time, the examination of [a] specific blade technology or parietal art and their occurrence or absence is not sufficient to define and to explain the behavioral capacities of *Homo sapiens.*" Behavior, you could say, is poorly constrained by the artifacts that play a role in it—you cannot predict the behavior from the artifacts, and you certainly cannot measure the skillfulness of the behavior by the complexity of the artifacts.

I've promised to offer a suppler conceptual matrix for thinking about how niche construction unfolds over time. At the center of this matrix is scaffolding, which allows us to think of a pattern of behavior not as something that is transmitted from one generation to the next (via direct observation, via finished artifacts) but that grows over time under determinate but plastic constraints. Scaffolding is not a new concept in the philosophy of evolution, but it has been absent from discussions of the key determinants of the emergence and disappearance of the kinds of cultural and cognitive flexibility needed to maintain continuity of cultural tradition in the face of rapid environmental change. In its absence, these discussions have gravitated toward that which is most salient in the archaeological and historical record: artifacts.

Artifacts do represent a genre of scaffold. In the last chapter I described how treadmills served as a scaffold for my own behavior, not to say my body. Running exclusively on treadmills for a period of years, I developed habits of posture and movement fitted to the treadmill. I experienced patterns of skeletomuscular development and joint wear specific to treadmill running. My sensitivities to movement, light, and sound, not to say pain signals of repetitive stress in my body, were attenuated in ways they would not have been had I been running in environments where I was constantly called upon to evaluate changes in the nonself world that did not originate from the movements of my body—pedestrians, buckled pavement, tree branches, slippery surfaces—and to anticipate the likely implications of these changes for my own movement. Treadmills were incorporated into my somatic niche, the constellation of capacities and constraints—sensorimotor, physiological, morphological, and environmental—by which I negotiated the everyday work of taking up space in the world.

Material artifacts can be scaffolds, but they are not the only kinds of scaffolds. Nor are artifact complexity and the complexity of associated craft traditions the only things that can be scaffolded. To expand and refine our sense of what can serve as a scaffold and what can be scaffolded, and to continue to develop an alternate vision of how culture, including but not limited to technology, mediates the relationship between a community and its environment over time, let's return to the Tasmania case and consider more systematically some of the oppositions alluded to above.

TRANSMISSION VERSUS REPRODUCTION

Above, I referred to the *social reproduction of skillful behavior* and implicitly contrasted *reproduction* with *transmission* as the term is used in many models of cultural evolution. *Transmission* evokes the dissemination of traits that exist in some stable, palpable form independent of the chain of evanescent events in which those traits are performed or, to use a term that will recur, enacted. Here again we see the influence of artifact analysis and, in a different way, that of population genetics. In artifact analysis, we can identify traits that inhere in artifacts and follow their dispersal and evolution over space and time. So, for instance, in a scraper industry such as that of the thumbnail scrapers observed in late Pleistocene assemblages in the limestone caves of upland southwest Tasmania, we could measure the length and breadth of individual scrapers, the degree of retouch they exhibit on their working surfaces, the materials they are made of, and so on, and try to establish trends in time and space. As we saw, in the case of thumbnail scrapers there are no clear trends. There is variation in size, degree of retouch, and material from site to site, but there are no clear lines of influence to suggest, say, that the community at one site had broken off from that at another site, taking with it a scraper tradition that it then developed in a novel way. In many cases, however, you can identify trends of this type in artifact traits. It is still imprecise to speak of the "transmission" of traits, since these traits are simply the material precipitates of a tradition of scraper manufacture. Excluding cases where the tradition was inferred and recreated strictly from the finished product (say, in the case of artifacts transported over a great distance, so that the recipients had no other knowledge of the makers), it is the tradition of manufacture whose dissemination you would like to observe.

One way to think about the distinction between artifacts with observable traits and a less easily observed tradition of manufacture that lies behind variation in these traits is by analogy to the distinction between phenotype and genotype. This is an analogy that feels natural to proponents of treadmill models. The analogy is not without merit. In particular, it encourages us to imagine a richer and more precise way of characterizing the process by which culture changes over time. But the analogy is also misleading insofar as it ignores the fact that the underlying traits are themselves habits of social behavior.

Like all social behavior, that implicated in the everyday activities of survival (making tools, getting food, building shelter, etc.) is reflexive and socially indexical. We are constantly observing ourselves and others and associating subtle variations in our ways of going about things with distinct social footings. Sometimes these footings are episodic and limited to a single interaction, sometimes they are more enduring. This lamination of behavior to footing is commonly observed in how we use language. Linguistic anthropologists refer to it as *enregisterment*. Registers are linked repertoires of palpable signs and strategies for arranging them—choice of sounds and manual gestures, choice of words, choice of grammatical constructs—that *index* or refer back to our role in a conversation or in the community. Often—most linguistic anthropologists would say always—the indexical content of a discourse, the meaning that arises from participants' use of language to comment on and reshape the social context in which that discourse unfolds, far outstrips the referential or propositional content, "what the words mean." Genes do not use gamete fusion as an occasion to comment on the social distribution of alleles. But people flag social distinctions in the use of language all the time. Not just that: often, when we use language, we are more interested in creating or maintaining a context for coordinate action than we are in communicating something specific.

The theory of registers originated with language, but it applies to social behavior more broadly. Social learning, whether or not it is accompanied by didactic speech (generally, in foraging communities, it is not), is not simply, nor even mainly, about the transmission of skills. It is, rather, about the recreation of a social world, which is why I feel it's more appropriate to speak of the social reproduction of skillful behavior than of the transmission of anything, be it artifact traits, knowledge, or skill.

Skill transmission models—those, including treadmill models, that begin with a capsule scenario of social learning—uniformly ignore the indexical dimension of communication. Instead, they start by formulating hypotheses about how would-be learners select a model, an individual whose behavior they will observe and imitate. In fact, to call these model-selection scenarios hypotheses is generous. Often they are quasideductive. That is, you start by abstracting from your own experience to imagine a set of alternatives—"Choose your mother (if you are a girl or young woman) or father (if you are a boy or young man) as your model," "Identify the most skilled person in the community and choose that person as your model," "Choose the highest-status person in the community," "Choose someone at random," "Don't rely on any one model, but adjust your behavior so that it conforms to what you observe among your peers." You select one alternative as the most plausible modal strategy, the way most would-be learners will find their model. To this model-selection strategy you add hypotheses about two distributions: that of skillfulness across the community at some hypothetical starting point—is everyone equally skilled, are skill levels random, is skillfulness normally distributed about a mean, etcetera—and that of individuals' learning outcomes relative to the skillfulness of their models.

The initial distribution of skillfulness tends to be relatively unimportant, since the distribution of learning outcomes rapidly restructures the distribution of skillfulness across the community. Indeed, debates about the validity of treadmill models, for mid-to-late–Holocene Tasmania or any other time and place, have tended to focus on scenarios for how an individual's skillfulness (or behavioral complexity, or fitness/adaptive value—as we've seen, skill transmission models confuse these) will diverge from that of her model, the person she's observed. These learning outcome distributions encode assumptions, generally tacit and untestable, about the looseness of fit between what you can observe of someone else's behavior and what you can infer about their distal intentions. Errors in the transmission of behavior from model to learner represent instances where a learner has misapprehended their model's intentions.

The question of how we infer what someone intends simply by observing and imitating what that individual does goes to the heart of what distinguishes social learning, and social cognition more broadly, in humans from homologous phenomena in other animals. This question has been the object of considerable laboratory research with infants, young children, and chimpanzees to elicit developmental and interspecific differences in styles of observation and imitation. It has guided a large body of work on the anatomy of the motor cortex and on the role of motoric entrainment—think of turn-taking in conversation or the more demanding forms of rhythmic phase alignment entailed in making music or dancing in an ensemble—in facilitating coordinate intention, coordinate action, and empathy or coordinate emotion. Much of the work in this field has been oriented toward identifying an explicit theory of mind by which we read others' intentions off their behavior. But among philosophers of mind there is growing conviction that you do not need an explicit theory of mind to explain how coordinate intention arises from observable behavior—it simply arises from an appropriately tuned tendency to motor resonance.

This is a theme we'll return to. Here, I wish simply to note that none of this is explicit, or even implicit but accounted for, in how skill transmission models select a distribution for learning outcomes. As for the other key parameter in skill transmission, learners' strategies for selecting a model: there has been less fine-grained observation of how learning unfolds in subsistence communities than you might imagine. But there is a small literature. Sometimes, though not always, skill transmission models refer to this literature to buttress a choice of model selection strategy. But the findings on real-world model selection strategies have been diverse. In some communities, or in some studies, young people seem to rely mainly on their parents for guidance in the acquisition of essential skills; in others they seek out additional models or look to peers. You can find evidence for whichever strategy you favor. Critiques of treadmill models have tended to focus on parameter values: Does the model show a decline in skillfulness over time when you simulate it with a "parents first" strategy or a "best in community" strategy? What happens if you make learning a two-stage process, so that at first learners look to their parents, then they seek out highly skilled or high-status individuals, adjusting their behavior accordingly? Less attention has been paid to the question of whether the transmission of skill is really the key phenomenon in social learning.

Again, there is something strangely disembodied about skill transmission models. We get just the vaguest picture of what observation and imitation look like on the episodic time horizon, the moment when learner and model are copresent, the one going about her work (or, less commonly, demonstrating what she's doing for the learner's benefit), the other observing and doing something similar. We get no justification for treating the observational learning of whole-body motor cascades (iconically, spear-throwing) as identical, for the purpose of modeling rates of change in community skillfulness, with fine-motor activities (say, plaiting fiber to make rope), skills that depend more on sensory than on motor discrimination, or on patterns of sensorimotor contingency that must be acquired more by experimentation than by observation (keeping a fire burning at an appropriate intensity), or with those that depend mainly on emotion regulation (cold tolerance). Transmission scenarios make no provision for practice, that is, for sensorimotoric and emotional regulatory rehearsal, whether with supervision from the model or at some remove from the observational learning context.

Even stranger, they make no provision for somatic limits to copresence. This becomes clear when you consider the implications of the treadmill model for Tasmania: had Tasmania not been separated from mainland Australia by the flooding of the Bassian sill, the pool of skillful models for observational learning would have remained much larger, and with it the richness and innovativeness of Tasmanian material culture. How would this have worked in practice? Would young Tasmanians have regularly sought out more skillful learning models from communities to the north? Would traffic in material innovations from the mainland have instigated a parallel wave of innovation in Tasmania? If the latter, what distinguishes social learning from artifact copying? If the former, when and how are learners and models coming into contact?

None of this is to say there are no demographic thresholds for cultural innovation. It is plausible at least to ask whether small communities are less likely to give rise to outliers, individuals whose behavior diverges from existing norms in potentially attractive ways. It is equally plausible to ask whether large communities encourage greater conformity, or whether both phenomena are active at the same time. But it is important, in this regard, to note that the selective constraints on innovation vary from one domain of behavior to another. So, for instance, in language, sound system may be subject to selection according to the topographic, physiographic, biophonic, and anthropophonic (human-generated sound) properties of the environment. Certain kinds of sounds and certain multisegmental phonological properties (pitch and stress contours over the course of an utterance or exchange of utterances) may be more perspicuous in places that are mountainous or flat, lushly or sparsely vegetated, quiet or filled with street noise. But the selective pressure that an acoustic environment exerts on the sound system of speech communities situated there will likely be relatively relaxed, allowing for a broad range of phonological niches to emerge. By contrast, we can expect topography to exert a more stringent degree of selection on locomotor style, with communities from mountainous areas adopting a walking style that emphasizes a low center of gravity and rapid adjustment to changes in terrain at the knees and people from steppes adopting a striding, hip-centered style that maximizes speed of displacement at the expense of responsiveness to terrain. Under relaxed selection, we should expect a lower demographic threshold for innovation.

Skill transmission models, lacking an explicit theory of how the environment, material and social, scaffolds behavior over time, cannot contend with variation in degree of selection and niche breadth. More broadly, they cannot contend with how the relative merits of skillfulness and technological complexity are conditioned by somatic coordination, both in the course of the exercise of skill (say, on a hunt) and in the subsequent redistribution of the goods gained by the exercise of skill (food, building materials, etc.). Anthropologist David Read offers an example:

> [A] single episode for attempting to obtain a seal through its breathing hole in the Arctic ice during the winter by a Netsilik hunter had high risk measured by the individual hunter's low probability of success combined with high cost of failure (because there were no alternatives to seal meat). However, the strategy could be implemented daily and by multiple hunters working in tandem, thereby reducing the probability aspect of risk when the probability of a success by any one of the multiple hunters is measured over the time period corresponding to the number of consumption days provided by the seal meat and blubber from the last seal that has been procured and distributed among the families living together.

So the coordinated exercise of skill, in tandem with strong norms of food redistribution, mitigated the cost of failure in an intrinsically high-risk environment. This returns us to the question of what role social learning plays in the life of the community—skill transmission? or social reproduction? Here at last I want to quote Henrich's observations of Tasmania:

> Among Fuegians [the peoples of what is today southernmost Chile, e.g. the Yámana, like the Tasmanians an iconic "most primitive" people], crafting bone-tipped arrows involved a 14-step process, seven different tools (four of which were specially crafted solely for making arrows), four types of wood (which all required straightening procedures), and six other materials. . . . In contrast, the Tasmanian technique of diving for crustaceans (which was exclusively women's work) probably requires both the development of substantial physical skills and lots of practice, but seems less likely to benefit from observing particularly skilled models. Such risky diving techniques are not, to my knowledge, used by other cold-climate foragers, and may have evolved in the absence of more complex food-procurement technologies.

There is so much going on in this passage. For one thing, gender bias: Henrich seems to be suggesting that the Tasmanian strategy is to be discounted by virtue of the fact that it was "women's work." For another, a comparison between a manufacturing practice and a food-gathering practice, as if these two kinds of practice entailed the same kinds of sensorimotor skills, were learned in the same way, served the same purpose in the productive life of the community, were subject to the same degree of selection and niche funneling, and could be scored against the same rubrics of complexity. He seems to be implying that honing locomotor-respiratory skill via practice (not to say the emotional control entailed in breath-hold diving) is less advanced, or at least less honorable, than identifying a highly skilled model to learn from—skill transmission models are strangely gerontocratic. We see an absence of comparative ethnographic curiosity (you would not have to spend much time trolling the literature to find that Korea and Japan are home to traditions of cold-water female subsistence breath-hold diving going back perhaps 2 ka). At the same time we see a discounting of innovation when it does not entail material elaboration—it is somehow a mark against the Tasmanians that breath-hold diving is *not* more widely attested as a foraging strategy. Finally, we see speculation that a deeply conserved (perhaps 3–4 ka) tradition of embodied skill serves as compensation for a lack of complex material technology, as if the latter were always to be preferred.

What we do not see is any concern with the social and somatic context of subsistence practice, whether manufacturing or food-gathering. Recall in the last chapter I mentioned Labillardière's observations of abalone diving (southeast coast, summer 1792 or 1793). The context is instructive. Out on foot with an exploratory party from the D'Entrecasteaux expedition, Labillardière came across a camp of Tasmanians, forty-eight by his count: ten men, fourteen women, twenty-four children, grouped around seven fires. Around midday, the women and girls went down to the water for abalone and lobster, surfacing to hand off what they'd gathered and caught to the men and to warm themselves at the fire before return-ing to the sea. This went on, in Labillardière's recollection, for some time. Surely the girls (if we take the Japanese and Korean traditions discussed below as ethnographic stand-ins, they would have been as young as eleven) were observing their mothers, other women, older girls, and one another, forming inferences about the intentions of those around them, and trying to reproduce both observable behav-ior and inferred intentions in their own bodies. But to characterize this scene as one of skill transmission is to miss the bigger picture.

As in practically all cases of social learning outside modern insti-tutions of schooling, learning here was embedded in the everyday productive life of the community. Indeed, though we don't have information about how much the younger divers contributed to the yield that day, we can say that learning was an integral part of pro-ductive life. Part of why the girls were in the water was to condi-tion their respiratory musculature, to improve their cold-water and pressure tolerance, and to sharpen their ability to spot things that were good to eat in the limited visibility of the undersea. But equally important was the indexical dimension of their joining the dive, the fact that by participating they affirmed both their own productive role in the community and the cooperative and redistributive nature of production. In situations where survival depends on cooperation and redistribution, rather than on the possession of a complex toolkit, we would do well not to underestimate the value of these indexical effects.

SURVIVAL VERSUS FLOURISHING

After writing the previous section, I was compelled to take a couple days off writing to attend to other things. I used the time away to improve my understanding of cold-water breath-hold diving. The facts suggest a further perspective on the difference between skill transmission and social reproduction.

Physiological research on breath-hold diving in northeast Asia dates to 1931, when the Japanese physiologist Gita Teruoka submitted an essay titled "Die Ama und ihre Arbeit" ("The ama and their work") to the journal *Arbeitsphysiologie* (*Occupational Physiology*). From around 1960, the *ama* and *haenyeo* ("sea women" in Japanese and Korean) began to attract more sustained attention from physical anthropologists and physiologists. In 1965 the US Office of Naval Research and the US National Academy of Sciences jointly sponsored a symposium on the theme in Tokyo. More recently, the growing popularity of competitive freediving has prompted renewed interest in the physiology of apnea (breath-hold) diving.

The basic ethnographic facts are these. In Japan, ama were found in communities along the Pacific coast from Chiba (east of Tokyo) as far south as Oita in Kyushu, and on the Japan Sea coast from the Oga Peninsula, in Akita, in the north, as far south as Nagasaki prefecture and the islands off the western coast of Kyushu. A cognate tradition of breath-hold diving existed to the south in Okinawa. Most ama dove part-time in the spring, summer, and fall, when the seasonal rhythm of subsistence agriculture allowed, though in some places ama dove in winter too. Some focused their efforts on kelps and other algae, others on gastropods and echinoderms. The details of how they dove varied. In some cases, they swam out to the dive site, tethered themselves to floating casks, and dove unassisted to depths of five to ten meters. In others, they used weights of up to twenty-five kilograms to allow themselves to descend to depths of fifteen to thirty-five meters. The weight was then pulled up by a male partner (a husband, brother, or father) stationed in a boat at the surface. Assisted divers could harvest more than fifty kilograms of algae, mollusks, and echinoderms per day. In Korea, haenyeo were concentrated in Jeju Island, where they dove year-round, swimming out or sometimes taking a boat to sites of five to ten meters' depth and using gourds for flotation. Census figures are approximate at best. One estimate put the total number of ama in Japan in 1949 at fifteen thousand, declining to seven thousand by 1963. In some parts of Japan men dove too, though their diving and harvest strategies differed from women's and their numbers were fewer.

In both countries, into the 1970s, divers wore nothing more than a cotton outfit, diving in water as cold as 10 degrees Celsius (the thermal conductance of water is more than twenty times that of air, with a correspondingly greater rate of heat loss from the skin); in some parts of Japan they dove unclothed. In summer, the diving day comprised three shifts of up to two hours, though at peak harvest times the workday could extend to ten hours; in Jeju, winter shifts could be as short as twenty minutes, with long periods in between warming oneself at a fire on shore (figure 3). The energy expenditure needed to compensate for heat loss was estimated at one thousand kilocalories for a day of diving, made up for by a greater food intake, relative to nondiving peers, in the morning and evening. By 1930, ama in Japan had begun adopting face masks. Prior to this they dove without masks, identifying edibles without benefit of corneal vision. In both traditions, dives were short—in unassisted methods, prior to the introduction of wetsuits, on the order of a minute, with half that time submerged and half resting at the surface. Girls would start diving at eleven or twelve, first in about a meter of water, attaining

Figure 3. Haenyeo warming themselves at a *bulteok* (enclosed hearth), Jeju Island, 2004 or 2005. By permission of the Jeju Haenyeo Museum.

status as fully competent divers at seventeen or eighteen. Some continued working into their seventies.

Physiologically, cold-water apnea diving represents one of the most extreme instances of somatic plasticity in the history of humanity. As you would expect, breath-hold divers exhibit enhanced vital capacity—the volume of air they can take in with a breath—and enhanced mobility in the thoracic cage. But these pale in comparison to the thermoregulatory conditioning they exhibit. Haenyeo working prior to the introduction of wetsuits in the late 1970s were found to have peripheral tissue insulative capacities superior to those recorded among Inuit. This was not due to subcutaneous fat deposits—in fact, haenyeo tended to be leaner than nondiving controls. Rather, they seem to have exhibited a combination of vasoconstrictive adaptation and enhanced brown adipose tissue thermogenesis of the type discussed in the last chapter.

More broadly, apnea diving presents formidable challenges both physiological and biomechanical. These include pressure: every ten meters of submersion below sea level introduces an additional quantity of pressure comparable to atmospheric pressure, so at twenty meters you experience three times the pressure on the thoracic wall and tympana (eardrums) that you would at the surface. Barotraumas of descent include pulmonary edema, alveolar bleeding, and atalectasis (lung collapse). These have been observed in single-bout dives as shallow as thirty meters, and evidence of pulmonary edema from prolonged surface swimming suggests that the repeated-dive pattern typical of ama and haenyeo would incur a heightened risk of edema even at relatively shallow depths. Ascent carries its own dangers, as the rapid depressurization of the lungs reduces the partial pressure of oxygen, creating a risk of hypoxia and loss of consciousness. Hypoxia of ascent is compounded by alternobaric vertigo, in which asymmetric changes in pressure in the middle ear between the two sides of the head can cause loss of awareness of one's rotational orientation.

Diving fasted, as was and is typical of haenyeo and ama, increases risk, as lipid metabolism is both less efficient than carbohydrate metabolism and generates less CO_2—you experience hypoxia faster, and your main chemoreceptive cue to breathe, carbon dioxide accumulation in the blood vessels and lungs, is depressed. Head submersion induces a combination of transient hypertension and bradycardia (slowed heart rhythm), combined, in cold water, with constriction of the airways—a pattern that ama and haenyeo experience up to 150 times a day. The risks of habitual diving, even to depths of no more than five or ten meters, include hypernitrogenemia (toxic nitrogen partial pressure) of the lungs, hearing loss, and tinnitus.

In Jeju, diving was regarded as a marginal occupation. In 1960, 40 percent of haenyeo were their family's primary earner. More than three-fifths said they would prefer to make their living some other way, and 64 percent said they would not allow their daughters to become haenyeo. Whether circumstances had been equally desperate in earlier eras is impossible to say. In 1960, Korea was still recovering from thirty-five years (1910–1945) of Japanese colonial rule and twenty-two years (1931–1953) of nearly continuous war. Still, looking at the Jeju case, we can identify an intuition, reasonable if shallow, underlying Henrich's dismissal of breath-hold diving in Tasmania: given the choice, most people would prefer to get their living in a less laborious, less risky way.

The question then becomes: What is laborious and what is risky? Labillardière and his contemporaries were impressed with the effort that Tasmanians, particularly the women, put into the work of securing food, but their accounts do not suggest Tasmanians were malnourished, nor that they found the food-gathering task oppressive. Comparative energy expenditure provides one rubric for ranking ways of life as more or less effortful, but energy expenditure is inadequate on its own. It tells us little about perceived effort, not to say satisfaction both in the exercise of skill (say, identifying and harvesting a range of targets sessile and motile without corneal vision under the biomechanical, thermal, and respiratory constraints of diving) and in coordinate effort.

These are the kinds of factors—perceived effort, satisfaction—that skill transmission modelers would dismiss as fluffy-headed conjecture with no material influence on the choices people make about how to keep themselves fed and sheltered. But this dismissal is unwarranted. What makes, say, satisfaction suspect is not that we have no evidence it plays a part in shaping how people act but that it is more difficult to operationalize—to assign scalar or even ordinal (rank) values to—than, say, energy expenditure or the number of parts in a tool.

I would go so far as to say that the difference between survival and flourishing is that the latter incorporates measures of perceived effort and satisfaction. Here we can start to see what is lacking in the theory of adaptation at work in skill transmission models. A theory of adaptation in the Darwinian sense—how circumstance shapes behavior to maximize reproductive fitness—will do a poor job characterizing change in human behavior over time. There are at least three reasons for this. One is that adaptation, as we've seen, is not simply a reaction to extrinsic change—it is a process of niche construction in which the community has a hand in shaping its environmental scaffold. A second reason why adaptation is conceptually brittle is that human life history is exceptional, among animals, for its elongated character, both during development and following reproductive senescence. Many of the selective cues that shape the human niche are most active in the postreproductive phase of life. The third reason to look beyond adaptation is that many of the cues that shape our behavior operate in the domain of flourishing rather than survival—the exceptional pleasure we take in participating in coordinate activity, for instance, or in the exercise of motor skill. Often these do contribute directly to a community's reproductive success, as in collaborative foraging and childrearing, but just as often they stand orthogonal to or in tension with reproduction.

Whether to track survival or flourishing is a value-laden decision. So is whether to treat change over time in culturally salient behavior as a process of adaptation or niche construction. These decisions overlap with, but are not identical to, decisions about what kinds of scaffolds we should expect to find when we examine adaptive behavior: technological? or enactive? *Enactive* is a new term, but it describes something we've observed repeatedly in this chapter and the last: situations in which the material residues of action are ephemeral but the action recurs and coalesces into enregistered patterns by virtue of how it becomes coupled to sensorimotor tendency. Critically, this coupling is dyadic: practice is a big part of how we transform a cascade of actions into embodied skill, but for something to really stick in the community, it needs to get passed back and forth between individuals—and your own performance, in the moment and over time, must become contingent on what you've learned to expect others to do.

Enactive scaffolding may seem precarious by comparison with technological scaffolding: What happens if the dyadic chain is broken and you have no durable material precipitates to fall back on to regenerate it? But, in fact, enactive scaffolding is the basis for language, music, and dance (and, the architectural theorist Bill Hillier has argued, for urban built form). For that matter, it is present even in cases that seem, on the surface, technological. You would be at pains to infer a foraging strategy from a pannier or a spear, certainly to the level of detail necessary to actually get food—that level of detail includes embodied skill that can only be acquired by enaction.

There is more to flourishing than adaptation, and there is more to the scaffolding of a secure niche than technology. When we view human responses to environmental change through the lens of technological adaptation, we limit our capacity both to identify flourishing in distant times and places and to imagine responses other than the technological to our own environmental crisis. More than that: we limit our capacity to identify dimensions of the response demanded of us by the environmental crisis that, like enhanced brown adipose tissue thermogenesis in cold-water breath-hold divers, do not exhibit a clear-cut projection of will into the material world the way artifact manufacture does.

Physiological adaptation is an expression of will in that it does not come about from a single episode of exposure to environmental stress. You have to go out in the cold, or the heat, or at altitude, or get in the water, over and over, every day, and, at the community level, over many generations. That willingness to expose oneself to environmental stressors when the relationship between exposure and acquiring something of value is labile and deferred is a mark of a distinctly human kind of cognitive flexibility. Its outcomes, somatic and environmental, represent a kind of emergent will. This is a theme we pursue in the next chapter.

3 Equilibria

One morning in 1997, when I was twenty-one, I found myself in the forest at four in the morning. This was June, winter in the Southern Hemisphere, and the forest in question lay on the Southern Ocean coast of Western Australia, near the old sawmill town of Pemberton. Southwest Australia has a winterwet climate. *Wet* here is a relative concept—as in other places so designated (the Cape region at the southern tip of Africa, Southern California, the Mediterranean periphery), the weather is dry year round, but winters tend to be cooler and damper. That year Australia was having its coldest winter in years, and I did not have enough warm clothing. I was behind on work and had got up early thinking I could use the quiet hours to catch up. It quickly became apparent I did not have the presence of mind for work. I was alert to nothing save the fact of how cold I was. Somehow I ended up outside. Two things have stayed with me. The first is the sky. Pemberton sits in the middle of a karri forest. Karri, *Eucalyptus diversicolor*, grow to eighty meters, and the forest around Pemberton is among the last places in the world where you can find that kind of late-seral canopy. At the time, there was little light pollution, or that is how I remember it. The sky was glaucous with stars, all of them unfamiliar, for I had been in the Southern Hemisphere all of ten days. The second thing that has stayed with me was the scent of woodsmoke. I could not describe for you any of the constellations I saw that morning. But even today, whenever it is damp and chill and there is the scent of woodsmoke in the air, I think of that morning, standing in the gravel drive outside a cold house in the woods, jetlagged and anxious about work, smelling the evidence of controlled combustion. There is something uniquely resonant about fire.

At a number of points in the last two chapters I alluded to Tasmanians' use of fire. I argued that the Tasmanian use of fire played out over at least three time horizons. These we could call the episodic, the biomic, and the somatic. The first corresponds to what we have in mind when we talk about the *use* of fire or any other thing: situations where human intervention in the world is guided by clear intent, situations where, if you asked someone, *What are you doing with that smoldering scrap of bark (scraper, snare, spatula)?*, they could respond coherently without giving it much thought. By and large, *use* in this sense is *episodic*. That is, it unfolds over durations and extents that we can take in as they happen and reflect on later roughly the way you would a scene in a film. Think of the different ways humans use fire. There is the use of fire to enhance the living environment: for warmth, for drying, for light, or as a smudge against insects. There is the use of fire for cooking. There is the use of fire in manufactures: to harden silcrete and wood, to distill tar from resinous barks, to smoke hides, to fire ceramics, to smelt ore, to smith metal. There is the use of fire in hunting, to smoke animals out of their burrows and vegetative shelters. There is the use of fire in ritual: as an object of contemplation, a tool for inducing hypnotic states. There is the use of fire in signaling at a distance. Even as the use of fire becomes mediated by additional technological apparatus, it remains episodic. Think of the use of fire for traction and locomotion, be it via steam, chambered combustion, or jet propulsion. Many of these uses will be familiar from the previous chapters.

Then there are ways of using fire that are *infraepisodic*—they unfold over horizons of time and space that are longer than what we can take in or reflect on as an event or scene. We've seen two of these. One is the use of fire in biome modification, to keep forests open and, by extension, to create a habitat for economically significant plants and animals. The other is the use of fire as a scaffold of physiological adaptation to cold. This last we saw among the haenyeo (female breath-hold divers) of Jeju Island as well as among the Tasmanians. These are uses of fire on the biomic and somatic horizons.

Sometimes, as in the case of warmth and cold adaptation or hunting and biome modification, the use of fire unfolds over multiple horizons simultaneously. Often it is difficult to say which purpose, and which horizon, were foremost in people's minds. This is not simply a product of the approximate nature of the archaeological record. Sometimes, even when you can ask people what they had in mind, it is unclear whether it makes sense to call the infraepisodic outcomes the products of *use* in the sense of goal-directed behavior. The people you talk to may acknowledge the long-term effects of setting fire to the land, for instance, or allowing a child to recover at a fire between bouts of cold exposure, but deny that these were part of their purpose in making fire.

Fire, then, offers a vantage point from which to consider the brittleness of the technological view of scaffolding—its failure, among other things, to account for ambiguities in how humans express purpose—and to formulate something better. This we will continue to do in this chapter. In this regard, the southwest of Australia offers a useful comparative case to Tasmania, differing both in climate and in historical circumstances in ways that allow us to expand our understanding of the conditions, enactive and technological, of human flourishing.

To begin to see the contrast in historical circumstances between Tasmania and Southwest Australia, consider the following remarks by Lieutenant H. W. Bunbury, who surveyed the part of the Indian Ocean coast of Western Australia known as Geographe Bay (south of present-day Perth) in 1836: "It cannot be denied that Western Australia, as far as it is known, is generally of a rather sandy barren nature, partly owing to the constant dryness and clearness of the atmosphere and climate and to the periodical extensive bush fires which, by destroying every two or three years the dead leaves, plants, sticks, fallen timber, etc. prevent most effectually the accumulation of any decayed vegetable deposit." In the absence of these fires,

> the labour and cost of clearing [the country] would be so greatly increased as to take away all the profit, and it would change the very nature of the country, depriving it of the grazing and pastoral advantages it now possesses. This has already been proved in the case of Van Diemen's Land [Tasmania], where, *in consequence of the transportation of the Natives to Flinders Island, and the consequent absence of extensive periodical fires, the bush has grown up thick to a most inconvenient degree,* spoiled the sheep runs and open pastures and afforded harbourage to snakes and other reptiles which are becoming yearly more numerous. *[My emphasis]*

This is a familiar pattern—the arrival of Europeans destabilizes long-established strategies of niche maintenance. In this respect, the story in Southwest Australia is similar to that in Tasmania, save that, as Bunbury's remarks suggest, the experience of the latter led some among the new arrivals to the Southwest to at least consider the value of conserving those prior strategies. (Bunbury goes on to note that Europeans might well try their hand at burning the land but would surely do a poor job of it compared to the indigenous people.) But there is another set of historical circumstances where the divergence between Tasmania and the Southwest is starker. These are the deep-historical circumstances.

There is a sense in which the archaeological record of this part of the world confounds our expectations about how the life of a community—or, let us say, a *population,* reserving judgment about the nature of the social structure by which that population is bound together—unfolds over time. To my ear, the concept of scaffolding, as I used it in the last chapter, entails an assumption of directional change over time, toward greater social and technological complexity (*development*) and greater energetic flux through the biosphere (*intensification*). Development and intensification are things we tend to take for granted when we reason about the time course of human history. When we do not observe them, we imagine something has gone wrong: that the population in question met with some catastrophic perturbance to its way of life (epidemic disease, colonization, a megatsunami of the type attested, in the Western Pacific, for the early Holocene) or that its way of life proved maladaptive, unstable, tending either to runaway devolution (Tasmania, in the hypotheses canvassed in the previous chapters) or niche depletion (perhaps our own fate). Could an absence either of development and intensification or of their opposites represent an achievement in itself? Or are steady states in history illusory, like a marble balanced on a saddle that divides two basins—stable in one dimension, unstable in the others, so that its stability depends on a transient absence of extrinsic change?

To address this question, we begin, again, with climate.

In chapter 1 I surveyed evidence about the recent climate history of what is now Tasmania. I alluded there to the difficulty of articulating evidence of change in temperature and humidity on a regional scale to corresponding evidence for the Earth as a whole. Glaciation and deglaciation, for instance, do not unfold uniformly across the Earth's surface, contrary to what terms such as "Last Glacial Maximum" would suggest. Even the direction of correlation between temperature and humidity differs from place to place and over time. By and large, the climate record has taught us to associate cooling trends with drying trends and warming with humidification. But the strength of this association—or, sometimes, its inverse—depends on a range of localized factors: topography, prevailing winds, vegetative cover, surface albedo, soil porosity. These, especially vegetative cover and, by extension, surface albedo, are themselves influenced by temperature and humidity, but they also have a certain inertia or path-dependence. Climate, that is, is nonlinear: it depends not just on present conditions but also on the continuing effects of past conditions. Further complicating matters, the past conditions in question unfold over a range of temporal and spatial horizons.

Evidence for changes in surface temperature and humidity on a millennial timescale comes, first, from marine sediment cores. Like the pollen cores considered earlier, these afford a more or less sequential archive of past events at the Earth's surface. Marine sediment cores consist predominantly of fossilized foraminifera, amoeba-like single-celled organisms that deposit a calcareous shell. The stable-isotope oxygen ratios in these shells bear witness to changing atmospheric and sea-surface temperatures, since heavier isotopes precipitate faster as the air cools. Foraminifera cores have been joined, in recent decades, by a range of genres of evidence with finer-grained resolution in the time domain, including tree rings, pollen cores, and ice cores.

Long-term trends in the Earth's climate have been driven by three kinds of *forcings*, extrinsic sources of change that are not themselves implicated in climate feedback loops. These include orbital forcing—change in the eccentricity of the Earth's orbit around the sun—together with axial precession (change in the orientation of the Earth's axis of rotation relative to the sun) and axial obliquity (change in the angle of the Earth's axis of rotation relative to the plane of the ecliptic, that is, the mean plane of its revolution around the sun). Orbital forcing is periodic, with cycles of 100 ka (eccentricity), 41 ka (precession), and 19–23 ka (obliquity). Periodic change in the intensity of solar radiation (solar forcing) and aperiodic change in volcanic activity at the Earth's surface (volcanic forcing) contribute to trends in climate over briefer spans of time.

Over the course of the Holocene, these three kinds of forcing have interacted with existing patterns of glaciation and vegetative cover to produce a climate history in three movements. First, precession and obliquity combined to yield an increase in incoming solar radiation in the Northern Hemisphere summer. The orbital phenomenon peaked at 11 ka, but its effects were damped until 9 ka by glacial cover in the Northern Hemisphere—an example of the nonlinearity alluded to above. From 9 ka, temperatures rose, driven by Northern Hemisphere summer insolation, peaking in different places between 6 and 5 ka. Since 6 ka, temperatures have been cooling and glaciers advancing in the Northern Hemisphere. Seasonality—the difference in temperatures between winter and summer—has declined too. In the past three hundred years, human activity has interrupted the cooling trend driven by orbital forcing, as anthropogenic contributions of CO_2, methane, and nitrous oxide to the atmosphere have sharply outpaced background trends—in the case of CO_2, yielding an increase in atmospheric concentration of more than a third.

As this account suggests, the climate record is strongly biased toward the Northern Hemisphere. The El Niño Southern Oscillation (ENSO) represents, as one review puts it, "the most important branch of internal variability of the global climate system," and the frequency of ENSO events has doubled over the past 6 ka, with implications for trends in humidity and typhoon and hurricane activity well beyond the southern Pacific. But the mode of coupling of fine-grained regional trends in temperature, humidity, and vegetative cover in the terrestrial Southern Hemisphere to the planetary trends documented in marine sediment cores remains poorly characterized. So as in chapter 1, we'll emphasize trends in vegetative cover as a proxy for trends in climate.

The vegetative regime of Southwest Australia is remarkable both for its diversity and its stability. Species diversity is an order of magnitude greater than that of the Southeast at comparable latitudes—some eight thousand species as against eight hundred—despite the fact that the Southwest is considerably drier. Diversity is linked to stability: the period since the Last Glacial Maximum has seen less turnover in vegetative regime in the Southwest than in the Southeast. In effect, the drier climate of the late-Pleistocene Southwest preselected biota for tolerance of the cooling-drying trend that marked the glacial maximum.

In fact, to characterize the climate that engendered the biota of the Holocene Southwest as a late-Pleistocene phenomenon may be to understate the stability of the southwestern meteorological regime. The plant fossil record suggests that a shift from temperate rain forest (broadly similar to what you see in present-day mid-latitude Chile) to dry-adapted open woodland and scrub similar to the pyrogenic biome we saw in Tasmania was underway prior to the onset of Pleistocene (marine isotope stage 2) cooling. It was driven, initially, by plate tectonic activity, as the Australian continent moved into its current position, where a seasonal high-pressure cell buffers the Southwest against the atmospheric water potential of westerly winds off the Indian Ocean. At the time depths in question, three to sixty million years, it is difficult to offer quantitative or even ordinal estimates of the relative abundance of different kinds of flora—the vagaries of deposition and tissue preservation are too great. So we cannot say with precision when dry-adapted families came to dominate the flora of the Southwest, simply that they were present—and, perhaps, formed a significant part of the regional flora—prior to the arrival of humans no later than 48 ka.

What are these dry-adapted flora? Again, we see overlap with the flora of the pyrogenic woodlands of the interior highlands of Tasmania: sclerophyllous (leathery-leaved) canopy trees, including eucalyptus and Casuarinaceae (ironwood), joined, in drier parts, by *Banksia* and multistemmed eucalypti whose scrubby habit is better adapted to low rainfall. Depending on soil characteristics and the incidence of fire, steppe or heath may dominate—grasses are more tolerant of sterile soils, that is, those that do not support the microbiota that provide nitrogen-fixing activity for the root systems of complex vascular plants. Again, these seemingly independent factors are coupled, as recurring fire fosters soil sterility—at length giving rise to the long-term dominance of grasses as in the prairies of North America and the heaths and peatlands of Scotland.

In the Holocene, the Southwest does evince a modest trend over time. On the southern coast, for instance, the palynological record shows *Melaleuca* (paperbark or tea tree, like those used by Tasmanians for bark-bundle canoes) dominant in swampy settings, succeeded, after 4.7 ka, by eucalyptus. The incidence of charcoal suggests a role for fire, certainly including anthropogenic fire, in this succession. Indeed, fire was so deeply implicated in the making of the modern Southwest vegetative regime that it does not make sense to think of it as a perturbance. Fire was, rather, a dimension of the biome, alongside climate, soil, and vegetation. Despite Bunbury's concerns, it does not appear that the frequency of fire has declined markedly since European colonization (though, perhaps, the relative frequency of anthropogenic and meteorogenic fire has changed). This is not to say that the period since colonization has not seen a change in vegetative regime. The clearing of land for agriculture and the introduction of livestock and Eurasian grasses, not to say the displacement of Indigenous communities and the suppression of their strategies of land management, including prescribed burning and the curation of surface water, herbivorous fauna, and extensive networks of cleared tracks, has brought about greater change in regional biome structure than that observed in the palynological record for the previous 11 ka.

So much for vegetative regime. What about the habits of the humans whose presence is attested in this part of the world from 48 ka? Here too, the secular trend, the directional change over time, while not absent, is exceptionally subtle. Archaeologist Joe Dortch and colleagues put it squarely: "There is little evidence for intensification" when you compare anthropogenic faunal assemblages from cave and rockshelter occupation episodes spanning the climate shift described above, from 40 ka, prior to the glacial maximum, to 0.4 ka, just prior to the arrival of Europeans. These assemblages come from limestone contexts—recall from chapter 1 the advantageous preservation properties of limestone—in what is now the Leeuwin-Naturaliste region, on the southernmost aspect of the Indian Ocean coast of Western Australia.

As always, the question of what should count as evidence of intensification is not without difficulties. In this case, the authors have adopted a prey-ranking strategy, using the body mass and mobility of different kinds of animals to estimate energetic return and difficulty of capture. As in Tasmania, we need to be cautious about assuming that small, mobile animals are intrinsically more difficult to capture—hunting tools made of organic materials, including snares and nets, do not fossilize the way spearpoints do, nor do the bones of birds and lagomorphs preserve as well as those of more skeletally robust animals. The archaeological record nudges us toward a ballistic bias in our fantasies about the foraging economy. Nor would a shift to smaller, faster (that is, "lower-ranked," hypothetically higher effort-for-yield) prey necessarily indicate that the population in question had depleted the large-bodied sessile macropods of the type we saw favored in Pleistocene Tasmania.

But to a degree, the faunal record in Southwest Australia is neutral with respect to the causes and signs of intensification, since what it shows is taxonomic continuity—the same kinds of animals appear in human debris over the entire span of Pleistocene and Holocene occupation. Two of the three dominant genera, *Isoodon* (bandicoot) and *Bettongia* (bettong, kangaroo rat) were preferred foods among the Noongar, the people who occupied this country at the time of European colonization.

This continuity of strategy stands in contrast to other parts of Australia, notably the Southeast and Cape York Peninsula in the northeast. There, peri-Holocene arid episodes prompted a diversification of resource base as the environment became less accommodating of existing human populations. With the return of a richer biota in the Holocene, this more diverse strategy gave rise to demographic intensification and associated innovations in social structure, including a less mobile way of life and, perhaps, persistent inequality. Across Australia, the Holocene sees innovation in lithic technology and, in the Southwest as well as the Southeast, the introduction of estuarine fish trapping. But in the Southwest, these innovations do not appear to be correlated either with growing population density or with dramatic changes in the mix of plant and animal resources that humans relied on for food.

Could the difference, again, be one of climate? Did the longstanding aridity of the Southwest preselect human subsistence strategy for the rigors of the glacial maximum just as it did the strategy (subsuming taxonomic mix alongside phenotypic plasticity) of the region's flora? Could the Southwest represent a kind of *antirefugium*, selecting for strategies that minimize the subsequent disruption of climate bottlenecks? This is a promising line of thought. But it is not borne out, at least on first inspection, by comparison with contemporaneous events in other winterwet regions. In the Cape region of Southwest Africa, you see a similar climate and a similar panel of vegetative resources to Southwest Australia. There, you do see evidence of resource base diversification in the millennia following the Last Glacial Maximum, followed, as the climate improved, by growth in population density and stylistic innovations in ornament and rock art that suggest cultural diversification. One difference between the two regions is that Southwest Africa is more rugged, with topographical barriers that might encourage symmetry-breaking, seemingly random divergences in strategy between similarly situated populations. The northern and eastern borders of the Mediterranean Sea offer another example of a winterwet climate that saw economic diversification at the end of the Pleistocene followed by intensification and development in the early Holocene. In the Levant, this process culminated in plant and animal domestication and the constellation of changes in economic strategy and social structure we associate with neolithization.

As with the Pleistocene cave sequences we reviewed in Tasmania, I must stress that the Leeuwin-Naturaliste sequences represent the work of a small number of individuals—bands, not chieftainships. Given the long stretches of time between successive layers in these sequences and the fact that some shelters feature Pleistocene sequences, others late-Holocene, but no single site includes a record for the entire span of time in question, it would be misleading to characterize these sequences as the trace of a single community, or even a single population. Absence of human debris, of course, is not evidence of absence of human presence. Clearly—*recall the interpretive challenge posed by scale, the need to exercise judgment*—there was continuity here, in toolkit and strategy and, if we venture a bit past the confines of the archaeological record, in ethos, in enregistered standards of perceived effort and satisfaction, in sensorimotor indices of discomfort and contentment. We need not be dismissive of the material constraints on well-being afforded by different kinds of environment, nor about the comparative advantages of an astringent but slowly changing environment as against one that is richer but more mutable, to imagine that something is at work here beyond the material exigencies of energetic balance—calories expended, calories ingested.

The interpretive challenge here is the converse of that we encountered in Tasmania. There, we saw discontinuity in the social reproduction of seemingly useful technologies but, if the accounts of Robinson and Jones provide any basis on which to reason, no apparent implications for the development of social structure. Here, we see not just continuity in the social reproduction of technology over a period of 40 ka, but also the incorporation of novel technologies introduced from elsewhere in the continent. We also see, faintly, a growing confidence in the use of fire to shape the land over the last 4 ka of the Holocene. Yet we do not see endogenous cultural development, either technological or social-structural, of the type observed under similarly stable if astringent environments in other parts of the world. In Tasmania, we see abiding environmental adversity— or what looks to outsiders like adversity—in the form of cold and wind. In the Southwest we see neither. And yet both places present instances of people "doing the wrong thing" from the point of view of trait transmission theory—either devolving maladaptively or failing to develop under appropriate stimulus.

Early on in the chapter, I introduced the image of a marble balanced on a saddle between two basins—stable in one dimension, unstable in the others. Unless you are endowed with unusual powers of visualization, when you imagine the marble you will see just one other relevant dimension—that which transects the saddle, so that we have, say, two hills, to the north and to the south, and two basins, to the east and to the west, with the marble suspended on the saddlepoint at the center. But I wrote *others*, plural, and I meant *others*. What I have asked you to imagine is a kind of *dynamic landscape* comprising not places but an arbitrary space of options, as, say, to trap fish or not trap fish, replant yam slips or not, wear clothing, shift camp on a regular basis, and so on. Before we return to this landscape view of adaptation, we need to say more about fire.

Here we run into old acquaintances. In 1968 Rhys Jones published a brief essay in *Australian Natural History* under the title "Fire-Stick Farming." Jones notes the regularity with which deliberately set fire appears in the ethnographic literature of Australia and catalogues purposes Australians might have had in setting fire: amusement, signaling, clearing paths, smoking out animals, stimulating the growth of economically significant plants, extending habitat. The first four correspond to episodic intents, those where the relationship between action and consequence is something you might infer from observing events as they happen. The last two correspond to what we'd call *planning*. That is, they entail an identification of causal relationships over intervals of time and space that exceed the horizon of episodic recall.

Now we start to see the relationship between fire and the forms of somatic cultivation discussed at the end of the previous chapter. In both cases there is a strange complementarity between the episodic motivation for creating a stressful environment and the long-term consequences. There are a number of ways we might interpret this complementarity. On the one hand, we could describe the long-term consequences—cold tolerance, a more productive environment—as positive spillovers or externalities. Or, if we favor the language of biology over that of economics, we might describe these long-term consequences as behavioral *exaptations,* instances where a phenomenon, having emerged under one set of selective cues (desire for abalone, desire for bandicoot), comes to serve another function altogether (inducing cold tolerance, promoting regeneration of a fire-tolerant biota with associated advantages). This first set of interpretations ascribes no awareness of the spillover benefit (exaptive function) to the people in question. Humans here are more the media than the agents of niche construction.

A second possibility would be to focus on the social-indexical dimension of episodic motivation. If you participate in some kind of stressful behavior (breath-hold diving, setting fire to undergrowth) because it is "what we do," perhaps the what-we-do-ness, the indexical component of the behavior, represents a kind of shadow planning—an awareness of the long-term beneficial effects, but an awareness that remains below the threshold of reflexive articulation. It might be that no one can say precisely why it's good to do *X*, but everyone is confident that it is.

A third possibility is that awareness of infraepisodic consequences is explicit. This awareness may or may not be linked to ascription of intent. That is, you might say something like: *Burning has known valuable long-term effects on the land, but our reason for burning is simply to clear a track (smoke out macropods, signal neighbors).* So even explicit awareness may not be enough to warrant the label *strategy* in the narrow sense of something done for the purpose of producing a stipulated set of outcomes.

The way I've laid out these interpretive options, they may appear to form an evolutionary implicational hierarchy, with the second following from the first and the third from the second. But I want to be cautious about projecting a process of dawning awareness upon the limited historical materials at hand. Equally, I must acknowledge limits to the analogy from somatic cultivation to biome modification. In past work I've argued that the two phenomena play coordinate roles in niche construction, or rather, that somatic cultivation should be understood as a dimension of niche construction no less than is biome modification—that the process by which we grow into a characteristic set of anatomical, posturokinematic, somatosensory, and metabolic dispositions is a value-laden one governed by a mix of conformance, accident, tinkering, and outright experimentation, just as is that by which we create the environment we inhabit. This still feels right to me, though there is a question here of the scope of the *we*, a point we'll return to in the final chapters. Still, the process by which we become aware of infraepisodic relationships between action and consequence—the process by which we develop a reflexive awareness of causation at scale—is different when the consequences unfold in our bodies as against out in the nonself world.

The difference is one of degree, not kind, but it is a difference nonetheless. It stems from the different roles that emotional coloring plays in how we assess the validity and significance of interoceptive sensations (as of stretch or ductile tension in muscles and connective tissues, distension in the gut, CO_2 partial pressure in the lungs) as opposed to exteroceptive and distal sensations—cutaneous tactile, visual, auditory, olfactory, and so on. To the interoceptive side one might also assimilate quasisensory sensations such as those of hunger, thirst, fatigue, sleep propensity, nausea, et cetera. The difference in the role of affect in shaping how we assess cues in these two broad categories is made clear by its absence in cases of *pain asymbolia*, a condition in which one experiences pain as a kind of tactile sensation absent the aversive quality we normally associate with pain. Ordinarily, emotional coloring shapes the experience of all interoceptive sensations, including, I suspect, the formation of inferences about the long-term consequences of episodic exposure to stress.

All this is simply by way of making clear that Jones's characterization of prescribed burning as *farming* is not incidental. Rather, it evinces a hypothesis about the nature of the causal reasoning—explicit, reflexive, deliberate—implicated in the role anthropogenic fire had played in shaping the sclerophyll woodlands and grasslands Europeans encountered when they first arrived in Australia. Jones was writing at a time when it was just becoming apparent that humans had been in Australia since prior to the Last Glacial Maximum. He was not the first to observe that fire was integral both to the character of the land and to the subsistence strategy of its longstanding inhabitants. But he may have been among the first to ascribe a willfulness to prescribed burning comparable to that usually reserved for the land-shaping activities of agrarian and industrial societies. The "promotion of regrowth through firing" observed in Australia, he writes, "is exactly the same process as that practiced by modern farmers burning off the stubble in a wheatfield, or by Welsh hill shepherds burning off the mountainside each winter to kill the old bracken. In all cases, *whatever the long-term effects may be* [emphasis mine], the immediate result of burning is to increase the quantity of edible plants for man and his beasts." That is, the ascription of will does not depend on a reflexive awareness of consequences over an arbitrarily long-term (say, transgenerational) horizon, but simply on a modicum of infraepisodic intent. In the contextualist spirit of chapter 1, it is worth noting again that Jones had grown up in Wales. He had, if not heard hill shepherds offer justifications for winter burning firsthand, absorbed those justifications in the conventionalized forms current in what was then an agrarian society.

Prescribed burning has been observed across Australia, and it has been inferred from the palynological, pedological (soil-derived), and vegetative record of Epipaleolithic and Mesolithic occupations in Europe and North America. But Southwest Australia is distinctive in that the clear evidence of the human use of fire, over hundreds of generations, to create and maintain a congenial environment is unaccompanied by any of the conventional markers of intensification. Should a pyrophytic biota be considered a marker of intensification in itself? That is, should we think of burning as the strategy the inhabitants of the Southwest adopted to increase the carrying capacity of their environment the way people in other places adopted the captive breeding of plants and animals, say, or, in wet areas, the manipulation of waterways to create padi? Was fire simply the instrument of intensification best suited to a winterwet climate, a high preponderance of sterile soils, and a biota poor in obvious targets of domestication? Or was it simply chance that the inhabitants of the Southwest, early on, perhaps prior to the onset of the Holocene, found themselves in an adaptive channel that allowed them to maintain a semblance of equilibrium, balanced at the edge of the basin leading into intensification, not to say a more laborious way of life, for so long?

The question of where to ascribe will to a pattern of behavior and where to ascribe contingency is value-laden, as was the question, in the last chapter, of which expressions of will to prioritize in assessing the adaptive felicity of one strategy or another. I do not wish to attribute unique ecological sensitivity or foresight to people in the deep past simply by virtue of the fact that they neither burned down their home nor went extinct. The ideology of environmental mastery that has come to serve us so poorly in our own day was not, initially, the cause of intensification but a consequence, a post hoc rationalization of a cascade of changes in economic strategy that unfolded with no long-term intent.

Could we imagine—*inductive generalization across scale again*—a regime of prescribed burning that underwent a similar process of development until the use of fire to manage the environment became as labor-intensive as the use of domesticated plants and animals? Or is this precisely what we cannot imagine with fire—precision, planning, mastery? You may have a precise understanding of seral succession, of how the vegetative cover and fauna of burned land will change over time following a burn. But you can never know, when you set the world on fire, exactly how the fire will spread, how far, how fast, in which direction, how hot it will burn. What makes fire such a powerful emblem of mastery, of order, is the fact that it is intrinsically aleatory. Fire is palpably random in a way other basic features of the material world are not. It violates the expectations of motor laminarity—smoothness, predictability—that we look for when we encounter some moving thing in the world and want to know if it is animate, living, if it is possessed of an interiority, a sense of its place in the world, that mirrors our own. Fire is alien, disturbing, and it remains so even when we have contained it and turned it to our purposes. Combustion, in fact, is the destruction of order, the reduction of organic matter to a less surprising, less information-rich state. Perhaps this is what makes the ease that Tasmanians and Noongar felt around fire, at the time of European colonization—not to say the similar ease evinced by present-day Indigenous fire managers in the Northern Territory—such a problem for theories of cultural evolution. It is as if Indigenous Australians have recognized affordances in fire, as if fire disclosed an inchoate tool-like potential to them, the way a hammer or a trapeze bar might to those from other backgrounds. To invite fire into your peripersonal space, to corporealize it, suggests not just a different historical trajectory but a different way of imagining the role of the body in making space.

At the time Jones wrote "Fire-Stick Farming" it was becoming clear that Mesolithic foragers in northern Europe and the British Isles had played a role in the vegetative successions of the terminal Pleistocene and early Holocene. In some places, open flowering woodland gives way to steppe-like grassland even as the climate grows warmer and more humid. In others, woodland comes to be dominated by pyrophytic genera such as hazel and birch. Charcoal horizons appear in the palynological record. Grazing (as opposed to browsing) fauna proliferate, helped along by the appearance of park-like grass plains bordering places of human settlement. None of this is proof that humans were setting fire with the explicit aim of instigating—or preventing—seral succession. These could, again, at least initially, have been side effects of activities undertaken for more immediate purposes: clearing paths, for instance, or driving prey—or even the products of calamity or malevolence, a camp fire gone out of control, a fire set to drive off rivals. Over time, as people observed the lasting effects of fire on vegetation and soil, and, by extension, fauna, weather, and climate, fire might undergo a kind of exaptation. At length, setting fire might come to be explicitly associated with the maintenance of a landscape and a society. This, archaeologist Gordon Noble hypothesizes, is what we see in Neolithic Scotland, where a ritual complex emerged centered on the burning of house-like timber structures. "The process of burning," Noble writes, "may have been the means by which memories of previous generations and past events were celebrated and retained."

We know remarkably little about the origins of controlled anthropogenic combustion. What we know comes principally from the dating of burned patches that co-occur with signs of human presence. If you find evidence of the combustion of vegetation in a discrete patch in or near a site of human occupation, and the radionuclide and optical luminescence dates line up, you assume the fire was a hearth. But could it not have been a shrub that grew up on the site soon after humans had abandoned it and that was ignited by a wildfire? Sometimes, if you look closely at the debris left by fire long ago and compare it to that left by contemporary fires, you can tell the difference between a burn mark left by recurring ignition over days or months and that left by a one-time ignition. The first could be a hearth; the second, who knows. All this assumes that the combustion debris has not been disturbed by the passage of time. At time depths on the order of a hundred thousand to a million years, this is not a safe assumption. So despite claims for controlled combustion at 1.6 million years ago, secure evidence that humans were *using* fire appears only after 400 ka. And—to return briefly to the theme of cold tolerance—clear evidence of habitual use of fire postdates human occupations of northern Eurasia by several hundred thousand years. At one point, humans flourished in cold-wet environments without making hearths.

Figure 4. Successive pulses from CP 1919 (now known as PSR B1919+21), 1968. Originally from Harold Craft, "Radio Observations of the Pulse Profiles and Dispersion Measures of Twelve Pulsars," PhD dissertation, Cornell University, 1970. Scan courtesy We Made This Ltd.

Over the past three chapters I have tried to trouble our assumptions about the relationship between technology and economic strategy. Indeed, I have sought to broaden our view of *strategy* to include patterned behavior that is not willful in the conventional sense—conducted with an awareness of scale—but rather the contingent product of a chain of gestures none of which anticipates the distal outcomes—just as, when we speak, we do not anticipate how our ephemeral gestures will contribute to the emergence of a novel speech register. Niche construction is not like running on a treadmill. It is, rather, like negotiating a rugged terrain.

If you are of a certain age, you can no doubt picture in your mind the cover art to Joy Division's 1979 album *Unknown Pleasures*. It features a reproduction of a plot of eighty successive periods of radio activity from pulsar CP 1919 (now known as PSR B1919+21), recorded by radio astronomer Harold Craft at Arecibo Observatory, Puerto Rico, in 1968. This plot offers a good way to start to think about dynamic landscapes. Each row of the plot shows a series of energetic peaks over time, concentrated in the window when the pulsar's radio pulse is visible from Earth. But from one period to the next, the exact procession of these energetic peaks varies. If you imagine the peaks as mountains and project yourself into the plot, you can imagine yourself picking a way down the plot, looking for places where the valleys between the mountains coincide from one period to the next. Now imagine you also had the capacity to inflect the disposition of peaks and valleys—not to shape the terrain from whole cloth, but to nudge it a bit so as to make it easier to pick your way from one row to the next. It is no accident that this thought experiment involves the skillful use of the body.

Let me further enrich the metaphor. Imagine now that instead of picking your way over a mountainous country on foot you have been in the cockpit of a giant anthropomorphic robot. The robot amplifies your actions, though often in ways you did not anticipate. It amplifies your capacity to shape the unfolding of the terrain over time while insulating you from the environment.

Now imagine the robot runs out of fuel and you must climb down and continue on foot. When you descend from the cockpit, what do you take with you?

4 Landscapes

Early in the Northern Hemisphere fall of 2017 my partner Jessy and I spent two weeks in Takashima municipality, Shiga prefecture, north of Kyoto. Shiga is a country of rolling hills covered in Japanese cedar and chestnut. It straddles the watershed that divides Lake Biwa in the east from the Japan Sea in the west. We had gone there to spend time off-grid, in a cabin built by a couple of back-to-the-landers who were raising their kids there, in a hamlet at the end of the sealed road. Yumie and Harufumi were about fifty. He was from Kyoto, she from Hiroshima, where her family had been resettled from Kyushu after the Second World War. In the 1980s, separately, they had become vegetarians for environmental reasons and had worked in the Indian restaurant scene in Kyoto, which is where they met. For Harufumi, vegetarianism led to environmental activism—protests against nuclear energy, against the deforestation of Sarawak, against U.S. petroadventurism in the Persian Gulf. At length they had moved to Shiga, wanting to give their children a different kind of grounding in the world.

They had four children, the third and fourth delivered at home in Shiga, the last when Yumie was forty-five, with no birth attendants. Their oldest son was in Kyoto, living at his grandparents', preparing for the university entrance exams. The next went to school in the nearest town, an hour's drive. The third boy walked twenty-five minutes to the elementary school, where he was one of three students. The following March, they told us, the youngest would join him, if she desired.

Our cabin lay a minute's walk from our hosts'. We slept on futons on tatami in a corner of the main room. In the mornings I would wake first, draw the kakebuton over my shoulders, and sit zazen facing the bookshelves that lined the wall at our heads. Most were guides to going back to the land: *How to Raise Strong Kids; The Half-Farming, Half-X Way of Life*, where X was written in romaji and bore its conventional algebraic sense: the thing you had to figure out. The bottom shelf was given over to blues and jazz biographies.

Our first day there it was raining lightly. Harufumi had drawn us a map and pointed out a waterfall at the source of the creek that ran by the house. In the rain we walked up the trail that ran along the bank, passing a dredging crew repairing the creek where a typhoon earlier in the season had left it silted in. On the way down, we discovered that our feet were covered in leeches.

We had been told autumn would be dry, but the monsoon had come a month late and now typhoon season was running a month behind schedule too. Torrential rain kept us indoors. We holed up and read, toasting brown rice to make genmaicha, making liberal use of the kerosene heater for hot showers. Scouring the breakfast pans with ash and water heated on the gas ring. I was reading about climate in Australia in the early Holocene—megatsunamis in Tasmania, the warmer, wetter conditions that prevailed eight thousand years ago. The second storm was a slower, soaking rain that lasted a full two days and left a chill in the air. The air had been thick with dragonflies—now it was still. Salamanders and praying mantises sunned themselves in the road, crows perched on the electrical wires, watchful.

The hamlet was called Onyuudani, "Small-Entrance Valley" (小入谷—the reading of 小, usually *shou*, is variant). In the afternoons we explored the surrounding area, sometimes accompanied by our hosts' dog, Miri.

We sensed, more than saw, a reddening in the hills. Mornings were cold. There was warmth in the sun but also an awareness that it was the clear days that were coldest by day's end. Harufumi gave us an old Coleman-style heater, its cream-colored enamel scabbed and peeling, the steel rusted underneath. My other reading that week was Cormac McCarthy's *The Road*. Lighting the stove in the mornings I thought of how McCarthy's protagonists say they are "carrying the fire." I marked this passage:

> When he woke again it was still dark but the rain had stopped. A smoky light out there in the valley. He rose and walked out along the ridge. A haze of fire that stretched for miles. He squatted and watched it. He could smell the smoke. He wet his finger and held it to the wind. When he rose and turned to go back the tarp was lit from within where the boy had wakened. Sited there in the darkness the frail blue shape of it looked like the pitch of some last venture at the edge of the world. Something all but unaccountable. And so it was.

I asked myself: If the world were ending, what would you take with you?

The third storm started mid-morning the Friday of our second week. It poured all day, rain slanting down, wind occasionally howling. Sometime between two and three in the morning it stopped and I heard insect stridulation. The morning was mild. Sun shone down through a clearing sky. In the afternoon we walked up the mountain road. Abruptly, the climate changed—it was drier, there was salt in the air, and the scent of the sea, and sea light, the last three spurring us to pick up the pace. The forest turned to pine. Looking out across the valley, you could see chestnut alternating with cedar on abutting slopes. Higher still, the microclimate alternated in the switchbacks, back and forth between dry and damp.

At the top, there was a lookout where you could see the Japan Sea—but on that day fog obscured the view.

The following day, we went to the neighboring hamlet, where friends of Yumie and Harufumi's were harvesting a small plot of rice by hand.

Two days later we returned to the world of internet and high-speed trains. In Kyoto, waiting for the Shinkansen to Tokyo, I opened the *New York Times* on my phone to the news of wildfires in Northern California. Seventy-seven thousand hectares (191,000 acres) consumed in five days. *If the world were ending . . .*

In the last chapter, fire emerged as a point where what I have called enactive and technological scaffolding meet. We saw how, in Southwest Australia as in Tasmania, humans used fire to shape land, to turn *environment* into the more specific *landscape*. I wanted to say that fire, though it is clearly technological in that it extends its manipulators' capacity to influence the world, falls on the enactive side of the distinction I had drawn in earlier chapters. Perhaps you were unconvinced, and, in a way, so was I. For really what I wanted to show was that the distinction itself, between enactive and technological, though helpful initially, is ultimately invidious. The distinction is helpful in that it allows us to see that not all adaptively significant components of a community's strategy for keeping body and soul together can be expected to be archaeologically salient. Nor—and this is more significant—can they be expected to yield to operationalization in the straightforward way we sometimes hope for when we are dealing with material things. If, in light of the discussion thus far, it now strikes you as somewhat shocking that serious people could treat mereological cardinalities—sum up the parts—as a proxy for technological sophistication, let alone for adaptive felicity—then, good.

But the distinction between enactive and technological is ultimately invidious because it suggests that there is a distinction of kind to be drawn between strategies that depend on how we use our bodies and strategies that depend on how we use bits of the world beyond the skin envelope. Toward the end of the last chapter I wrote, *It is as if Indigenous Australians have recognized affordances in fire, as if fire disclosed an inchoate tool-like potential to them, the way a hammer or a trapeze bar might to those from other backgrounds. To invite fire into your peripersonal space, to corporealize it, suggests not just a different historical trajectory but a different way of imagining the role of the body in making space.* This phenomenon of corporealization, of inviting the world in, is not so rare as my words might suggest. Indeed, corporealization is pervasive. Think of how, drawing a sharpened pencil across paper or fabric, you feel the surface texture of the material as if your body ended not at the tips of your fingers but at the tip of the instrument.

Even so, we can identify different tendencies, divergent ethoses perhaps, in strategies of making land. These we might label, usefully if invidiously, enactive and technological. And these, I suggested, correspond to different kinds of outcomes in a conceptual landscape of survival and flourishing, what we might call the parameter or feature space of adaptation.

In this chapter we continue our discussion of the relationship between biospheric space and the feature space of community adaptation, shifting now to focus more on the feature space. In this way we turn to the contemporary world and our collective future.

As we will see, to speak of divergent *outcomes* in an adaptive landscape is misleading, for it suggests, as does the notion of an adaptive landscape itself, that the feature space of adaptation is more or less fixed and that our task is to seek out stable optima in that space. In fact, the space of options is itself plastic. It changes under extrinsic forcings such as the orbital processes described in the last chapter. It also changes under our own actions. Again: evolution, cultural and epigenetic no less than genetic, is a process of niche construction, in the space of possible strategies just as in the physical space of the biosphere.

As the vignette with which I opened the chapter suggests, I am concerned, in this latter part of this book, with the relationship between how we negotiate topography—say, walking a watershed, be it Shiga or, as in *The Road,* Appalachia—and how we negotiate the topology of the feature space that influences, among other things, what typhoons and wildfires will be like a couple generations from now. What is perhaps less obvious is that I am equally concerned with privilege. To spend time in a place such as Onyuudani or any quiet, remote place is increasingly a mark of privilege—economic, educational, cultural. Indeed, to adopt a more pared-back way of life, whether for a week or for good, to play at asceticism, or to take it seriously, is an option available to a thin slice of the Earth's human population. It is tempting to imagine our own present adaptive basin as a valley with a narrow entrance—or exit. But there are two ways of thinking about how the way out is narrow. One is strategic: there is a narrow range of strategies by which we might escape our rapidly disintegrating niche, and they all demand a certain astringency of behavior. The other way of thinking about what it means for the way out to be narrow is in terms of privilege: If the way out is narrow, who gets to go through?

In an elegant 2017 essay in *Philosophical Transactions of the Royal Society B* (that is, the series of the *Transactions* dedicated to the life sciences and the social sciences), the evolutionary biologist Douglas Erwin offers a capsule history of the use of the *space* and *landscape* metaphors to describe evolutionary and adaptive processes. He also discusses the limitations of these metaphors as they've been elaborated thus far. Briefly, the origin of the *space* metaphor is widely attributed to the geneticist Sewell Wright, who in 1932 imagined gene loci as the dimensions of such a space and the various alleles of those genes as points along those dimensions. A decade later, in *The Major Features of Evolution* (1944), the paleontologist George Simpson extended Wright's concept to that of an adaptive landscape by the addition of a scalar measure of fitness, of the type I alluded to in chapter 1—total fertility at the individual or community level, say. Fitness runs orthogonal to all the feature dimensions of the space—if you imagine a space of two features, as is common in efforts to reason about evolutionary spaces, it is easy to project these two dimensions on those of the Earth's surface and fitness on elevation. Thus, a landscape.

In passing, note that in many, perhaps all contexts, as we've observed, scalar measures of fitness are inadequate. But Erwin points to a raft of more fundamental challenges for the space/landscape metaphor. Many of these stem from our tendency to use low-dimensional spaces, and above all those with Euclidean or spherical geometry, as sources of intuition for reasoning about the behavior of the high-dimensional spaces of evolution. The landscape metaphor has lulled us into believing that evolutionary spaces are topologically similar to the spaces of our physical world. But this belief is unwarranted. In some cases, evolutionary spaces may be *pretopological*. That is, every point in the space might have a set of neighboring points, and a chain of such neighbor relations might allow us to trace out a path through the space—but the nature of the neighbor relation might not conform to any uniform metric or distance, whether Euclidean or otherwise. In other cases, we might observe spaces with a meaningful, uniform distance function, but distance might not correspond to *accessibility*, the probability that an evolutionary search will transition from a given point to one of its neighbors. Consider, with Erwin, the space of RNA conformation: "The folding of small RNA molecules has provided a powerful tool for understanding evolutionary topology, and particularly the relationship between genotype (the RNA sequence) and the three-dimensional structure produced by folding of the RNA sequence (the phenotype). . . . Since many changes to a sequence will not generate a change in the folded structures, there will be large networks of effectively neutral changes. *Single base-pair changes to the RNA sequence may trace a path through the sequence space, but single base-pair changes can also produce very different optimal two-dimensional structures.*" That is, "some random networks will have many adjacencies to other random networks," while "other networks will be relatively isolated, with few adjacencies to other neutral networks." While the sequence space is endowed with a roughly Hamming metric (i.e., counting any one-glyph change in a string of glyphs as a topological distance of one—this entails that all single-nucleotide substitutions are equally likely and the sequence length remains constant, though in fact neither of these assumptions is valid), the space of three-dimensional shapes evoked by those sequences is pretopological.

This is a trivial demonstration of the difficulty of mapping the evolution of complex behavioral phenomena, be it RNA conformation or collective subsistence strategy, onto change in some underlying medium of transmission. Indeed, the RNA example is a bit too simple to guide our thinking about change in collective behavior because in the RNA case, unlike in cultural evolution, three-dimensional conformations do not exert selective cues "upstream" on underlying sequences—the flow of selective cues is one-way. In the case of behavior at the level of organisms and communities, including human communities, matters are otherwise. It is not clear that it even makes sense to describe the evolutionary spaces of the different dimensions of human behavior, however you want to decompose them—genetic, epigenetic, sensorimotor, metabolic, symbolic—as laid out in an implicational hierarchy, sequence, or layer cake, upstream to downstream, bottom to top. It is rather that they form a dense network with selective cues flowing bidirectionally between all pairs. Indeed, it is more than this. We must also include nodes in our network for various dimensions of change in the environment—that is, the evolutionary process is one not of adaptation but of niche construction. And we must take into account the fact that often we observe multiple communities of living things with unrelated germlines and patterns of behavior evolving as metacladal unities or *holobionts*—think of the relationship between humans and the microbiota that colonize our orifices, skin, and gut.

Erwin notes drily: "Since roughly the same number of genes (18,000 to 20,000) is needed to make any animal, the *complexity of animals reflects not variation in sequence space but in the regulatory interactions that enable development.* Thus, the various spaces are not necessarily decomposable to a sequence space."

The multiplicity of the topology of evolution is one challenge we must address if we are to benefit from the landscape metaphor. A second is the fact that the spaces of evolution change, as do the qualities of "diffusion" or "search" through them. Erwin offers the example of the origin of the animal clade:

> Molecular clock studies indicate that the origin of Metazoa occurred approximately 780 Ma [million years ago] although animals do not appear in the fossil record until approximately 560 Ma.... Although some specific genes are related, animal genomes are about two times larger than their ancestors; many proteins include novel domains or arrangements of domains; regulatory networks have evolved through the construction of new types of circuits; and the range of operators that change regulatory interactions has expanded, as with co-option of subcircuits within a GRN [gene regulatory network]. *Developmental spaces are discontinuous because of the introduction of the means of cellular coordination, the introduction of distal enhancers near the origin of Metazoa, and the generation of entirely novel developmental spaces via signaling pathways and microRNAs.*

In the domain of culturally transmitted behavior things are, if anything, more complex. Not only does encultured behavior direct cues to the epigenetic, genetic, and environmental nodes in our network of corresponding landscapes, it also, as it were, interferes in its own evolution via the social-indexical processes we canvassed in chapter 2.

The causal multiplexity of evolutionary cueing makes it difficult to set out the changes that mark the human niche today, some of them alluded to in the opening vignette, in an expository way. There is no single best place to start. But let's give it a shot. In what follows I outline some of the key phenomena that archaeologists five thousand years from now, be they recognizably human or otherwise, will attribute to a turn in the evolutionary trajectory of humans and our cohabitants. These come under two broad headings, urbanization and climate change.

URBANIZATION AND ITS CLINICAL SEQUELAE

Part of what makes places such as Onyuudani—or, in a different way, the post-apocalyptic Appalachian Piedmont of *The Road*—so compelling is that they offer a contrast to the urban spaces that have, over the past generation, become the modal environment of humans and a number of other living things. *Clinical sequelae* is a strong way to put the implications urbanization has had, in particular, for mammalian and avian populations. Is it warranted?

The stressors that urban life exposes animals to include air- and water-borne pollutants, noise, and, perhaps surprisingly, an excess of stressful social contact. Chemical and aerosol pollution now represent, as a recent report by the Lancet Commission on Pollution and Health puts it, "the largest environmental cause of disease and premature death" among humans. Some 16 percent of deaths are attributable to pollution in some form or another, a vastly greater proportion than that attributable to infectious disease, armed conflict, or everyday violence. You might be inclined to think that pollution, certainly air- and water-borne pollution, by virtue of its ambient nature, is a great equalizer. It is true, according to World Health Organization estimates, that nine in ten humans in the world are consistently exposed to fine particulate aerosol concentrations that exceed the WHO's own safe exposure thresholds. In some places, notably sub-Saharan Africa and the eastern Mediterranean, rural households are exposed at higher rates than urban ones. This may reflect a widespread process of periurbanization in which increasingly dense but not-quite-urban populations continue to rely on charcoal for cooking and heating fires. Or it may reflect the siting of pollution-generating industries in rural areas, as with concentrated animal-feeding operations and oil refineries in the United States. In most parts of the world, cities rely, by design or happenstance, on surrounding areas to serve as pollution sinks. Still, it is city dwellers themselves who bear the brunt of the burden of chemical and aerosol pollution. This burden is unevenly distributed. In high-income states, urban inhabitants have a one in two chance of being consistently exposed to danger-ous levels of aerosol content. In low- and middle-income states, the chance is more like nineteen in twenty.

Aerosol pollution is easy to grasp, presenting, as it does, a mechani-cal insult to breathing and vision, not to say olfaction or the sense of smell and, at sufficient intensities, cutaneous tactile discrimination in the extremities. But this is just one class of insult among many. The neurotoxic and mutagenic effects of heavy metals (as lead and sulfur) and the endocrine-disrupting activity of persistent organic compounds are, if anything, more troubling.

Let's consider endocrine disruptors. These present two, perhaps three, distinct kinds of stresses to the individual. The first is that they insert themselves into hormonal signaling pathways, includ-ing those mediating sex and gender expression, metabolism and growth, and the emotional evaluation of social encounters. They can serve as agonists for endogenous steroid hormone-signaling mol-ecules, binding steroid hormone receptors in ways that mimic the endogenous signal, thus amplifying the physiological cascade along that signaling pathway. They can also serve as antagonists, binding receptors in ways that do not participate in the signaling pathway but prevent endogenous signals from participating. In some cases, the same molecule may serve as both agonist and antagonist in the same pathway depending on its concentration, giving rising to a *non-monotonic dose response*.

Endocrine-disrupting chemicals also interfere in epigenetic sig-naling, and this in two ways. First, they disrupt epigenetic modula-tion of gene expression (for instance, via methylation and histone modification) at specific loci in the genome. Second, they disrupt the cellular apparatuses of epigenetic modulation itself, including that implicated in the conservation of epigenetic markings in the genome from one generation to the next. This last phenomenon remains obscure, in part because the process by which epigenetic markings are propagated through the germline is poorly understood. This process entails *developmental reprogramming*, a genome-wide clearing away of epigenetic markers at the zygote stage, followed by their renewed application during cell differentiation. In primordial germ cells—gamete precursors—a second round of deprogramming occurs prior to final differentiation into germ cells, where some epigenetic markings reappear to be transmitted in reproduction.

Thus there are multiple points at which events at one node of our network of topologies, the environmental, say, can influence events at another node, in this case the epigenetic.

Exposure to acoustic noise provides a further illustration of how selective cues flow in all directions between environment, behavior, physiology, and epigenetic regulation of the genome. *Noise* can refer to a number of different things. It can mean sound at pressure levels noxious to the organs of hearing. Or it can mean sound that is lacking in the appropriate spectral structure, with energy spread across the audible frequency spectrum rather than concentrated in the frequency bands significant to humans, such as those that correspond to the formants of human speech. Or it can mean simply sound, even if it is well structured and relatively quiet, that is contextually unwelcome, as the yap of the neighbors' dogs when you are writing. All of these things are stressful, though the combination of high sound-pressure level and low spectral structure is especially so. Exposure to sound-pressure levels in excess of 90 decibels—common in many bars and restaurants—causes mechanical trauma to the hair cells of the basilar membrane, with one hour of exposure to 105 decibels sound sufficient to cause a permanent threshold shift in hair-cell activation in the frequency bands of exposure—that is, hearing loss. Recently it has come to light that prolonged or recurring exposure to sound at pressure levels long considered mechanically safe—below 80 decibels—can yield irreversible changes in audition by depressing excitatory synaptic connectivity for the frequencies of exposure, both in the inner hair cells and in the auditory cortex.

But the effects of noise on our capacity to cope with our environment extend beyond the auditory domain. These other-than-auditory effects are mediated in large measure by how recurring exposure to sound at high sound-pressure levels impinges on the autonomic nervous system and the hypothalamic-pituitary-adrenal (HPA) axis. In both cases, the initial step in an animal's homeostatic response to sound is the release of catecholamines and glucocorticoids— excitatory neurotransmitters and stress hormones—into the blood- stream. Chronically elevated levels of glucocorticoids contribute to impaired cardiovascular function and heightened risk of myocardial infarction and stroke. Not surprisingly, chronic noise exposure has also been associated with impaired cognition, notably in children. Across a broad range of animal taxa—arthropods, seahorses, finfish, birds, mammals—chronic acoustic stress has been found to nega- tively affect reproductive rate, development (including incubation in birds), metabolic efficiency, cardiovascular health, sleep, cogni- tion, and immune function. It is widely hypothesized that long-term modulation of neuroendocrine signaling in the HPA axis mediates these effects.

Indeed, chronic exposure to stressors of all kinds—acoustic, chem- ical, nutritional, social—becomes inscribed in the HPA axis via epi- genetic regulation of glucocorticoid and brain-derived neurotrophic factor (BDNF) expression. Under chronic stress exposure, genes associated with glucocorticoid and BDNF pathways (that is, genes for proteins forming parts of the receptor as well as the signaling molecules themselves) undergo methylation of upstream promoter regions: they are epigenetically modulated for increased expression. This modulation appears to be developmentally sensitive, so that a stressful childhood will have lifelong consequences in the form of heightened stress response and, via high BDNF expression, height- ened risk of depression. These effects appear to be hereditable.

Perhaps the most surprising risk associated with urban living is schizophrenia. Again, the nature of the effect implicates a window of heightened sensitivity during development, as individuals born and raised in densely populated areas show increased incidence of schizophrenia even when they do not spend their adult lives in cities. The effect even exhibits a kind of dose response: you are at greater risk if you live in a larger city than in a smaller one, and you are at greater risk the longer you spend living in cities. Though the correlation was observed as early as the 1930s, its mediating principles have eluded characterization. Again, the HPA axis appears to be implicated, along with what is known as danger-associated molecular patterns or DAMPs, a signaling cascade that evokes inflammation in the absence of tissue trauma or pathogens. Here, the precipitating stressor seems to be social in nature, with those who experience a higher rate of negative social encounters, notably migrants and the disenfranchised, at the highest risk. Evidence for the joint role of social appraisal and stress response comes from a combination of the systematic appraisal of the environmental correlates of clinical presentation ("deep phenotyping"—keeping track not just of an individual's complaints and symptoms but of where and how they spend their days) and functional brain imaging. Most intriguing of all, spending time in green spaces, even simply parks within urban areas, lowers the risk.

Thus far we've concentrated on one pathway by which selective cues flow from the environment to behavior, that mediated by a neuroendocrine stress response that both induces changes in behavior episodically and exerts a ratcheting effect on sensitivity to stressors via epigenetic marking. The episodic response amounts to a turning up of sensorimotor gain. Think of your own fight-or-flight response: it is as if the whole world, including sensations and actions originating from within your body, gets louder. There are other pathways, intertwined with but distinct from the neuroendocrine, that have yet to receive the same degree of attention from urban ecologists. We'll consider just one, oxidative stress.

Oxidative stress refers to the body's burden of reactive oxygen and nitrogen species with the potential to oxidize, and thus de-structure, molecular constituents of the body's tissues and signaling apparatus. More precisely, oxidative stress is the molar ratio of reactive oxygen and nitrogen species to endogenous antioxidants. The latter include electron donors that quench the oxidative reactivity of the reactive species, together with more elaborate enzymatic apparatus for isolating reactive ions and guarding the body from them. The avian ecologist Caroline Isaksson has hypothesized that exposure to oxidative pollutants might reduce an animal's capacity to cope with sensory and social stress via the action of glutathione (GSH). GSH serves both as a neurotransmitter and as an endogenous antioxidant. Its neurological role appears to be largely anxiolytic. In this hypothesis, enhanced oxidative stress would vitiate the individual's capacity to modulate its response to other kinds of stressors such as noise, crowding, or intimidation by diverting GSH from its role in the central nervous system (modulating autonomic response to sensory and social stressors) to address the oxidative burden of nitrogenous and metal-ion pollutants.

As Isaksson's hypothesis suggests, how birds—or humans—respond to the stressors of a novel environment depends in large measure on how those stressors co-occur. This makes it difficult to extrapolate from observations of how animals respond to the same stressors when they occur separately. The challenge of the urban environment is *multimodal*.

Our response to that challenge is equally multimodal, and it unfolds over at least three distinct time horizons simultaneously. In the last chapter I referred to three horizons over which the use of fire as a tool of niche construction played out: the episodic, the biomic, and the somatic. The three horizons I have in mind here are related. We could think of them as the episodic ("coping behavior"), the developmental-lifecourse ("acclimatization"), and the transgenerational. This last is known in the ecology literature as adaptation. Given the problems we've seen with the use of this term to describe change in behavior, we need a richer language. Perhaps again *somatic horizon* will do.

For most living things, urbanization might as well as be an extrinsic forcing similar to the orbital and geological phenomena we looked at in chapter 3. For humans, by contrast, it feels incomplete to speak of urbanization as something that simply happened to us.

Have we not brought it about through our own actions? For all the stressors of urban life, there must be selective cues driving humans to spend more and more of their time in densely populated environments—opportunities for waged labor, for instance, not to say social stimulation. The same could be said of the pressures nudging us to spend more and more time online. These cues operate over a relatively short time horizon—if not the episodic then certainly the everyday. They match our imaginative templates for how a typical day should go, or perhaps what we might accomplish in the next few years. By contrast, the cumulative effects of the stressors of urban life are slower to materialize. We are less likely to associate these effects with episodic instances of the stressors that precipitate them. So the selective pressure they exert on our reflexive behavior, the part of our behavior that draws on our capacity for causal inference, is weaker.

CLIMATE CHANGE

Urbanization represents one key dimension of the environmental scaffold of somatic behavior today. Climate change represents another. For all the talk about what climate change portends—for our capacity to feed ourselves, for sea levels and, by extension, the habitability of present-day coastal zones, for the incidence of catastrophic bushfire and rain, for the susceptibility of humans and other living things to infectious disease—there has been remarkably little discussion of what it portends for the moment-to-moment sensorimotor experience that makes up the tissue of our day. Will it be uncomfortable in the ordinary sense in which, say, a hot, humid day is uncomfortable? Will it induce similar sensations of enervation, a depletion of our drives to move, eat, or do much of anything? Will it demand—or induce—an emotional conditioning in the young comparable to what we saw when we looked at cold tolerance in chapters 1 and 2? Will it have neuroendocrine and epigenetic follow-on effects similar to what we saw above? How would you begin to formulate hypotheses on such questions?

In a pair of elegant studies, the political scientist Nick Obradovich and colleagues propose that we can, at least, formulate hypotheses about the effects of climate change on motor tone and circadian rhythms of activity and rest by extrapolating from what we know about how people respond to extreme heat and humidity today.

The behavioral data come from an instrument known as the Behavioral Risk Factor Surveillance System (BRFSS), implemented by the US Centers for Disease Control and Prevention. The BRFSS originated in 1984. It is conducted continuously, rather than in the periodic "waves" more common with surveys. Essentially it is a random telephone survey, though in some places in-home interviews figure too. At this writing, the Centers for Disease Control, in cooperation with state and territory public health services, conducts on the order of four hundred thousand BRFSS interviews every year. The interview takes approximately twenty-seven minutes. It covers a lot of things: chronic conditions, immunization, tobacco and alcohol use, cancer screening, HIV, mental health, exercise, sleep, seatbelt use. Optional modules that state and territory health authorities may include at their discretion cover respiratory health, sun exposure, cognitive decline, and respondents' roles as caregivers for others. Many of the questions take the form, *In the past twelve months, have you . . . ?* or *In the past thirty days, did you . . . ?*—that is, they rely on *retrospective recall*.

Unless you spend a lot of time reading social science survey reports, this may surprise you. Retrospective recall may seem like a remarkably inaccurate, not to say dated, way of finding out what people do. Is it not possible that people's recall will be distorted—either by their belief that they should have acted other than they did or by a desire to present their best self to the interviewer or by simple failure of memory? Yes, of course, all these things are possible. But it is also possible to formulate questions in such as way as to minimize the risk that these problems will erode the reliability of the data. In the case of exercise, for instance, the survey asks, *During the past month, other than your regular job, did you participate in any physical activities or exercises such as running, calisthenics, golf, gardening, or walking for exercise?* Notice that the question does not ask the respondent to estimate how often they work out or even how many times in the past month—the question is formulated to elicit a response along the lines of *yes, no,* or *don't know/not sure.* This may seem like a steep cost to pay to minimize the risk of inflated responses—wouldn't we prefer to know how often people work out rather than simply whether or not they have in the past month? But pooled over hundreds of thousands of responses, information-thin questions of this type can generate a textured picture of behavior as it unfolds over the population, if not as it unfolds in the lives of individual survey respondents.

Would it not be better to use some form of continuous surveillance that does not depend on participants' capacities to evaluate their own behavior from the standpoint of a neutral third party? In many places, notably the United States, health insurers have begun to pressure subscribers to adopt activity-tracking devices with an eye toward formulating finer-grained rubrics of risk behavior. As I've discussed in earlier work, these devices do not offer anything like the objective characterization of behavior their proponents would have you believe, nor is it clear that the data they provide offer a clearer picture of how people act than, say, retrospective recall.

For all the difficulties it entails, often the best way to find out what people do is simply to ask them. Even the recent techniques of so-called *deep phenotyping* alluded to in the discussion of schizophrenia above rely mainly on established techniques of anamnesis and physical diagnosis, together with the collation of education and hospitalization records. For all the hype surrounding technologies of self-tracking, continuous remote sensing has, up to this point, played a rather modest role in public health surveillance. This is not to say its role will not increase over the coming generation. It certainly will—encouraged, no doubt, by the success that a handful of states, notably South Korea, had in adapting mobile phone location signals to contact tracing during the Covid-19 pandemic that began in 2020—though more often, I suspect, it will do so in ways not conducive to flourishing. My own view, based on my experience designing prototype systems for the remote sensing of mood, alertness, and social stress, is that the richest way to incorporate remote sensing into social science is by using it to enhance the degree to which participants' capacity to reflect on their own states of being gets folded into the data. This reflexive capacity is not the problem it is often made out to be. The greater problem is the methodological blindness that comes from imagining we could devise value-neutral measures of behavior.

So, operating on the assumption that BRFSS survey data offer a reliable if limited view of how adults in the United States behave, what do we observe if we correlate the weather at the time when the survey was administered to a particular individual with that individual's recall of behavior over the preceding thirty days? Using 1.9 million responses to the "physical activities or exercises" question over an eleven-year period, Obradovich and Fowler conclude that there is a non-monotonic relationship between temperature extremes and recreational physical activity. That is, both extreme lows and highs of temperature tend to discourage people from being active. Precipitation also has a modest discouraging effect, though humidity yields no correlation. They conclude that if temperature trends continue on their current trajectory, nonoccupational physical activity in the United States will see a net increase by the end of the present century. This net increase masks considerable geographic and seasonal variance, with summers seeing a marked decrease in activity across the survey region and the South seeing a decrease year-round.

This may not sound like a very informative result, if by informative we mean unexpected. What makes this study elegant is how it takes a dimension of existing survey data that has, by and large, been treated as incidental—to wit, the time of year at which a particular interview was conducted—and shows how, paired with unrelated data, in this case meteorological data collected outside the context of the BRFSS, that previously uninteresting dimension can become interesting. At the same time, studies such as this one are limited in the same way as the laboratory studies alluded to above in our discussion of how the stresses of urban living affect birds and other living things: you cannot estimate the effect of a combination of cooccurring phenomena simply by summing up the effects of the constituents considered one by one. Again, the challenge posed by climate change, like that posed by urbanization, is multimodal. More precisely, it is *metamodal*—that is, it transcends the challenge posed by any one constituent.

In the case of physical activity, there are additional wrinkles to consider. For one thing, globally, over the past two decades, physical activity levels have declined in ways that suggest not so much an effect of urbanization per se as an effect of specific features of the urban environment as we have come to know it. The authors of a recent study on "worldwide trends in insufficient physical activity" speculate that increased use of "personal motorised transportation" might account for part of the marked increase, between 2001 and 2016, in the proportion of adults in high-income countries not getting enough movement over the course of the day. At the same time, East and Southeast Asia, despite rapid urbanization, have seen a decline in the proportion of individuals with insufficient physical activity levels, a trend attributable in part to the rise of recreational sports and a wellness culture in urban China. Worldwide, gender represented a significant determinant of risk of inadequate physical activity, with women in many parts of the world less likely to travel outside the home for work and less likely to engage in vigorous recreational activity.

What these results suggest is something we've seen all along: if you want to understand how culturally significant behavior evolves in response to changes in the environment, you cannot start by taking culture to be epiphenomenal to some underlying phenomenon arising from the interaction of environmental features and economic necessity. Nor can you usefully separate those parts of a behavioral strategy that are value-neutral (survival) from those that are value-laden (flourishing).

Perhaps nowhere is this clearer than in the domain of sleep. Concerns about a long-term decline in the number of hours of rest we get at night are not new. But over the past ten years sleep has become the object of a proliferating array of technical interventions, while adequate rest has become an emblem of class privilege. Rest, like physical activity, is subject to environmental factors. Obradovich and colleagues, again working with BRFSS responses, this time to the question, *During the past thirty days, for about how many days have you felt you did not get enough rest or sleep?*, demonstrate a striking correlation between temperature extremes and incidence of self-reported underrestedness. Their results suggest that, in the absence of adaptation, by the end of the present century, on average Americans will be underslept about one night in two—though the effect is more than three times as great in low-income communities. Note that this is the projected impact on sleep of changes just in climate (or, more specifically, the incidence of nighttime temperatures in ranges that would be anomalously high by present-day standards). This says nothing about the continued erosion of sleep by other factors, including those associated with urbanization.

The key phrase in the last paragraph is *in the absence of adaptation*. Of course we can expect to see adaptation, and again, over multiple time horizons and with difficult-to-foresee tradeoffs. Children born into a warmer world will almost certainly find it easier to sleep on warmer nights, at least up to a point. Some of this tolerance will almost certainly find expression in epigenetic modulation of the endocrine signals implicated in the circadian oscillation of glucose tolerance (that is, capacity to absorb food energy), sleep propensity, core body temperature, and desire for social contact, among other factors. (If you want to become more aware of these rhythms in your own body, observe yourself closely in the days following a long transmeridian flight, when your endogenous pacemakers are desynchronized from environmental rhythms of light and socialization. You will find yourself, for instance, inexplicably cold in late afternoon, when it is the middle of the night for your body.) Some of this epigenetic modulation may be transmitted to future generations.

Indeed, one recent hypothesis about the evolutionary significance of human sleep holds that humans are distinguished by the versatility of their sleep behavior. Humans, it has been observed, sleep significantly less than other primates, and our sleep habits are more variable. Within populations, you see a divergence of *chronotype* between individuals who, stably over the course of a lifetime, are more active early in the day and those who are more active late in the day. Between populations, you see a divergence of sleep strategies. These range from segmented sleep characterized by a relatively short nighttime sleep episode with a high incidence of brief periods of wakefulness, complemented by daytime napping, to the consolidated or "monophasic" sleep pattern characteristic of North Atlantic societies since industrialization (at least up to the introduction of the mobile phone). Human sleep is also characterized by an exceptionally high degree of REM sleep and a more rapid cycling between slow-wave and REM phases of sleep, perhaps corresponding to the elaboration of episodic memory in humans. Under laboratory conditions intended to reproduce the conditions of midlatitude preindustrial winter nighttimes (fourteen-hour dark periods), the sleep of participants acculturated to sleeping in a single consolidated nighttime sleep episode rapidly shifts to the biphasic pattern attested for the early modern era. The human circadian rhythm is itself remarkably resilient, capable of maintaining entrainment to a twenty-four-hour period even under the attenuated environmental cues of the polar regions. It would be difficult to say that any of these conditions (save those of the laboratory studies) is intrinsically inimical or conducive to human flourishing. Or rather: contextually salient rubrics of flourishing incorporate contextually salient strategies of sleep.

None of this is to say that human sleep is endlessly plastic. It is not, and the physiological and emotional toll of sleep deprivation is significant and growing. Climate change will no doubt contribute to that growth—in ways that most disadvantage those with the least control over their environment.

BORO: SOCIAL COMPLEXITY

Recall my comment, at the end of chapter 1, on the challenge of inductive generalization across scales of social complexity. One early reader of this book expressed concern that my treatment of urbanization and climate change is a bit glib. The crux of this reader's reservations concerned the sizes of the populations involved—or rather, the social complexity entailed in social reproduction on the scales implicated, say, in urbanization and climate change. In light of the themes of chapters 1 and 2, this should come as no surprise. Population size matters—but how, exactly?

The evolution of social complexity—how we get from populations like those of pericontact Tasmania and Leeuwin-Naturaliste to populations like those, say, of Tokyo, remains an area of vigorous disagreement in anthropology and evolutionary biology. There are at least three going camps, three ways of characterizing social complexity in humans as an evolutionary phenomenon, which we may designate by the terms *ultrasociality*, *group selection*, and *inclusive fitness*. All three camps represent efforts to address the same underlying question: *How could altruism—an individual's expending energy that it could use to support its own survival and reproduction on the survival and reproduction of others—evolve?* How, that is, could helping behavior, let alone the forms of coordination and cooperation we see in humans, get a toehold in a community of individuals that, at the outset, have little metabolic or reproductive incentive to cooperate?

If you're reading closely, you'll note that that last sentence slips in a prior assumption—that, "at the outset," individuals have little incentive to cooperate. Is this assumption justified? That is, in the absence of some specific mechanism to explain the emergence of cooperation, *must biological individuality, defined in terms of auto-poiesis or some related concept, entail a Hobbesian war of all against all?* This prior question is one that rarely gets asked in the literature on the evolution of altruism. We'll return to it at the end of the present section.

Ultrasociality has become the preferred term of art among economically inclined biologists and economists seeking to establish evolutionary foundations for their discipline. Definitions of the term vary, but a key criterion is that ultrasocial societies get their living in ways that depend on systems of permanent, coordinated, often coercive division of labor such as those observed in eusocial insects (wasps, ants) on the one hand and agrarian—agricultural, food-producing, as opposed to foraging—societies on the other. Recently, ultrasociality has been adopted by some ecologists as a framework for making sense of the origins and growth of what ecologist Erle Ellis has referred to as anthropogenic biomes or *anthromes*. It is Ellis's use of ultrasociality, among other things, that the early reader alluded to above had in mind.

Ellis's concept of the anthrome is one that I have found resonant, but I find the attribution of anthropogenic biomes to ultrasociality unconvincing. In fact, in my view no one has yet produced a robust account of the intermediate or mesoscalar structure of human sociality that would bridge the forms of niche construction we looked at in chapters 1–3 and those we've looked at in the present chapter. But some accounts are more convincing than others. In what follows I will try to explain why.

One common way of describing ultrasociality is with reference to the *permanent, coordinated, often coercive division of labor*. But something else that is common, perhaps universal, in discussions of ultrasociality is its proponents' interest in bringing the forms of sociality observed in eusocial insects and those observed in some human societies into a common framework. To students of the history of sociobiology and its successors, this will come as no surprise. But this conceptual unification turns out to be infelicitous, because the demographic conditions in which enduring coordination has evolved in social insects are different from those in which it has evolved in humans.

In a careful review of the mechanisms by which recurring reciprocal helping behavior might become enregistered in a community of living things, the theorists of eusociality Simon Powers and Laurent Lehmann argue that the key divergence is in strategies of reproduction. In social insects, the number of individuals in a colony that actually reproduce—the "queens"—is low relative to the size of the colony: generally one or a handful in a colony of thousands. Thus the degree of relatedness among members of a colony remains more or less constant even as the colony grows, and the risk that expending effort to help another member of the colony will hurt one's own reproductive fitness—say, by conducing to that other's having offspring that would compete with one's own offspring for resources—remains low. This is because related individuals share reproductive fitness—the propagation of the germline of one is (partly) coextensive with the propagation of the germline of the other. Similar conditions obtain in microbial communities, where reproduction is largely clonal in nature (even when you factor in ubiquitous horizontal gene transfer). Thus, the *indirect benefit* of reciprocal helping behavior—the fact that expending part of one's own finite resources to help others redounds positively to one's own reproductive fitness—goes a long way toward explaining how other-directed behavior could become fixed in a community over evolutionary time (compare the "somatic horizon" discussed above).

If this is the first time you've given thought to the conditions in which recurring other-directed behavior might evolve, it's not unlikely you're getting stuck on the question of anthropomorphic interiority: How does the ant, let alone the amoeba or bacillus, *know* that helping, if reciprocated with such-and-such a frequency, will have a net beneficial effect on its reproductive success? This is a

common confusion in the reception of models of *inclusive fitness*—
that is, models that allow for a gradient benefit arising from behavior directed at the survival of individuals other than oneself on the
basis of the relatedness of helper and helped. The confusion has
been exacerbated by the fact that inclusive fitness is often referred
to as *kin selection*, suggesting a kind of tribalism at work—making
decisions about whom to help on the basis of who is most closely
related to you—and prompting demands for the identification of
mechanisms (pheromone signaling, etc.) by which organisms might
"know" their kin. In fact, no such mechanisms are necessary, nor is
any kind of discrimination about whom to help that corresponds to
outward signs of relatedness. It is enough that a community exhibit
a certain spatial viscosity over time—that an individual is likely to be
surrounded by individuals that share, indirectly, via relatedness, in
its own reproductive fate.

In humans, as in many large animals, things are otherwise—
relatedness declines rapidly as community size increases, spatial
viscosity may or may not be a reasonable assumption—so inclusive
fitness is not enough on its own to establish cooperation as a dimension of a community's survival strategy. It is here that you need to ask
how individuals estimate the benefits of helping behavior. In some
cases, these benefits will be realized at the time of the exchange or
through repeated interactions between the same pair of individuals.
More realistic models of large human societies must also consider
benefits realized through chains of assistance—*A* helps *B*, *B* helps *C*,
and so on until at some point someone helps *A*—and more diffuse
forms of "generalized reciprocity"—I condition my expectation that
I will realize a benefit from helping future partners on the degree
to which my own past partners have reciprocated my help either
directly or via chains of indirection, even if I know nothing about the
actual past behavior of prospective partners. Indirect and generalized reciprocity both entail concepts of *responsiveness* (tendency to
reciprocate), *reputation* (publicly available information about individuals' responsiveness), and *norms of cooperativeness* (the statistical

distribution of responsiveness across a community). By incorporating these concepts into models of how individuals behave in a community, we can construct simulations that offer hints about the conditions under which cooperativeness will take hold in a community over time. If cooperation is to spread, initially there must be some net benefit to the cooperative subgroup—"productivity," "fitness," or what have you must increase faster than the number of individuals behaving cooperatively. But, as Powers and Lehmann note, "it is hard to find actual empirical cases where group productivity increases linearly with total investment into helping"—above a certain level of cooperativeness the selective pressure to become more cooperative relaxes, so the relationship between group size and degree of cooperativeness will likely be sigmoidal or s-shaped, increasingly slowly at first, then more rapidly, then more slowly again.

So far we've been speaking as if responsiveness—tendency to cooperate—along with the faculties of social cognition needed to keep track of the reputations of those in one's social network, were transmitted in the germline. But of course in humans norms of cooperation, like other dimensions of behavior, are transmitted socially, as are symbolic scaffolds for the scaling-up of social cognition, among them the forms of classificatory kinship found in Australia. So now we are back to the question that occupied us in chapter 2: How does social learning style (e.g., conformist bias, vertical versus horizontal transmission) interact with survival and flourishing? Here again, a combination of simulation and empirical evidence gives cause for skepticism about the heuristic models of social learning common in the trait-transmission literature—we should expect, or at least be prepared to reckon with, "marked interpersonal differences in the social learning strategies that different individuals use in the same setting."

In passing, I must refer briefly to the long-running, often acrimonious debate between proponents of inclusive fitness and those of group selection, an alternate framework for thinking about the evolution of cooperativeness. The questions of what it would mean for selection to act on groups (for instance, how and where transmissible traits would be represented, especially in forms of life where social behavior does not dominate the phenotype) and whether there are aspects of cooperation that cannot be explained by inclusive fitness alone have occupied evolutionary biology since the 1960s. In practice, the two formalisms are equivalent in their explanatory power. The differences between them are not ontological but epistemological—or, if you prefer, aesthetic. There are certain phenomena that some thinkers find it more natural to imagine with reference to selection acting on the group "as a whole" rather than on a community of interacting individuals. The evolutionary ecologist E. G. Leigh offers as a criterion for distinguishing the two situations the degree to which the fates of different constituents of the community are coupled, so that factors influencing the fate of the group eclipse those influencing the fates of its constituents. He offers endosymbiosis an example of such a situation, with organelles such as mitochondria and chloroplasts the constituent individuals whose fates have become entrained to that of the group. We can imagine boundary situations where our view of the appropriate formalism flickers back and forth much as it does in response to illusions you might have seen as a child such as the Necker cube, Rubin's vase, or the studies in multistable figure-ground symmetry of M. C. Escher.

A more interesting problem for going accounts of the evolution of social complexity concerns the cognitive challenge for an individual of keeping track of relationships as the social network expands. Most popularizing discussions of this cognitive challenge begin and end with *Dunbar's number*, after the anthropologist Robin Dunbar, who has proposed that the number of personal relationships a primate individual can keep track of scales with neocortex size (measured in brain mass, complexity of synaptic network, or what have you), with humans limited to approximately 150 such relationships.

The challenges of operationalizing a concept as rich, diffuse, and polyvalent as social cognition are many. As we saw with cultural complexity and adaptive fit in chapters 1 and 2, the case that it is ontologically desirable—that is, that it yields some kind of useful

parsimony of explanatory factors—let alone methodologically felic-
itous to reduce such a concept to a scalar magnitude (number of
"technological units," number of offspring, number of relationships)
turns out, on inspection, to be weak. There is, you could say, no harm
in trying, just to see what you get. I do not have the space here to
do justice to the questions of what kinds of evidence you'd want to
adduce to a scalar operationalization of social cognition, nor what
rubrics should guide inductive generalization from that evidence
to a hypothesized universal of human cognition—let alone the evidence you'd then need to test that hypothesis. But consider simply
the question of how to define *relationship*—and whether it is even
appropriate to bring all relationships under a single heading for the
purpose of gauging social cognition. I defer to the anthropologists
Jan de Ruiter, Gavin Weston, and Stephen Lyon: "The term *relationship* is a heuristic tool for understanding what is happening socially
between embodied agents. The grey areas as to what constitutes an
agent in this context are profuse. Dunbar's idea of a relationship,
which we could refer to as a 'grooming relationship,' prioritizes the
face-to-face contact of agents." But grooming, a category formulated
on the basis of observation of other-than-human primates, represents just one subset of a broader class of what de Ruiter and colleagues call *transactional relationships*. These vary not just in the
body-dyadic dynamics of "social bookkeeping" but in the valences,
that is to say, the qualities of emotional appraisal, we ascribe to
them: "In blood feuds and other rivalries, the knowledge of one's
enemies' kinship and friendship circles is as important as the knowledge of one's own. Do humans also have a limit of 150 enemies?"

Above I alluded to the role of systems of classificatory kinship—
roughly, totem affiliation, though I've tried to avoid the term *totem-
ism*, as it comes with unhelpful baggage—as scaffolds for scaling up
our capacity to manage relationship networks. De Ruiter and colleagues point to other such scaffolds that play a complementary role,
allowing us, as it were, to flip from an inclusive fitness to a group
selection view, seeing the group as the entity by which individuals

are constituted rather than the reverse. These two forms of scaffolding are complementary. Among the fusional devices enumerated by De Ruiter and colleagues are a number—national sporting events, communal violence—that work by evoking in participants variants on what the early sociologist Émile Durkheim, drawing mainly on published field observations from Indigenous Australia, called *collective effervescence*. Collective effervescence remains, more than a hundred years since Durkheim first introduced the term, hotly debated—What is it? How to operationalize it? What physiological correlates does it have? What cognitive demands does it make of participants, and what cognitive faculties, for instance, the distinctive human facility for picking up and entraining to a driving rhythm, were selected for together with it? What seems, to me at least, less debatable is that neocortical constraints on social network management are not what matter for formulating generalizations about social complexity. What does matter? Again we are back to a value-laden question about the ontology of evolution: What should we make salient in our models? In my view, we must give more emphasis to a range of archaeologically illegible and difficult-to-operationalize forms of somatic scaffolding—the enactive artifacts I've been harping on since chapter 1.

In one fashion or another, most of the authors whose work I've canvassed in this section recognize this. Powers and Lehmann, in their discussion of what is missing from trait-transmission theories of cultural evolution that would be necessary to explain the emergence of complex societies, lay great emphasis on *institutions*, metastrategies that afford interactants a way of renegotiating norms of interaction over time. "More formally, an institution is a set of game forms . . . a (communication) mechanism whose outcomes are rules for social interactions"—they offer as an example the allocation of shared resources for the policing of common-pool goods, for instance a timber reserve, that are susceptible of overexploitation by noncooperating members of a community. "The set of all possible allocations then corresponds to the set of game forms"—say, how much time each household is expected to contribute to guarding the forest, or how much money they are expected to contribute so that the community may hire a guardian, not to say what proportion of fines imposed on violators are granted to the guardian as a way of deterring collusion. By incorporating traits for "inclination to participate in the allocation institution" and "inclination to allocate a given proportion of shared resources to the maintenance of the common-pool resource" into their model, they are able to show that institutions, at least under this formalization, can become widespread in a population of interacting agents even if, initially, the vast majority of interactants are disinclined to participate in the institution.

This is a start. What is missing is a picture of how the metagame, the inclination toward institution negotiation, might itself emerge from a cascade of dyadic encounters within a community. We cannot simply imagine it as a binary trait whose value is susceptible to mutation, "asocial" to "social," in the course of individual acts of biological or cultural reproduction. This, perhaps, is the crux of what I glossed over in the earlier draft of this chapter—the intermediate scale, or rather, the chain of intermediate scales, between dyadic sense-making and biome modification on a planetary scale. Here it might help to ask what the relationship is between institutions, as Powers and Lehmann use the term, and registers as I've used the term in this book, that is, as an extension of the concept of speech register as it has been used in linguistic anthropology. One of the advantages of the speech register concept is that it allows you, in principle if not always in practice, to trace institutions all the way down the discourse chain to the individual encounter. This level of detail is generally omitted from simulations of the type alluded to above, but there is no reason (say, limitations of computational power to run the simulation) one could not formulate a model that took as its unit of iteration not the generation but something like the conversational speaking turn. This would require model-builders to work closely with anthropologists of a more particularist, and—to return to the disciplinary schism alluded to in the preface—interpretive-constructivist as opposed to analytic-reductivist bent. This type of cooperation is long overdue.

More broadly, as a field of research, the evolution of complex societies is overdue for a large dose of fine-grained historicism. In its absence, the field has grown thick with broad-brush conjectural history, often with a curiously hypermasculine tang. This phenomenon is related to the question I posed at the top of the section: *Does biological individuality entail a Hobbesian war of all against all* that may, under certain conditions, yield to cooperativeness and complex sociality? Why must the community be epiphenomenal to the interactions of individuals, with the individuals metaphysically prior? What if, instead, we took an enactive approach, starting from the premise that individuality is contingent on some kind of dyadic or collective phenomenon—that individuals emerge from the social plasma like the persistent moving figures that emerge, under certain conditions, in a reaction-diffusion process (figure 5)? How might our models of the evolution of social complexity, not to say our ways of imagining adaptive landscapes, look then?

An enactive approach to modeling the evolution of social complexity would pose new problems of formalization, but not, I suspect, insurmountable ones. Perhaps, as with inclusive fitness and group selection, there would be no difference in formal descriptive power between enactive and individual-first models. Perhaps, rather, there would be an epistemological difference, a difference in what felt salient—not to say what felt possible.

The first three chapters of this book represented an argument about the nature of the scaffolds of culturally significant behavior. *Scaffolds*, recall, are the phenomena that both support and constrain the development of some further phenomenon over time.

In its crudest form, the questions at the center of those first three chapters were, *Are the scaffolds of culturally significant behavior predominantly technological in the conventional sense, that is, mediated by the durable material precipitates of economic strategy,*

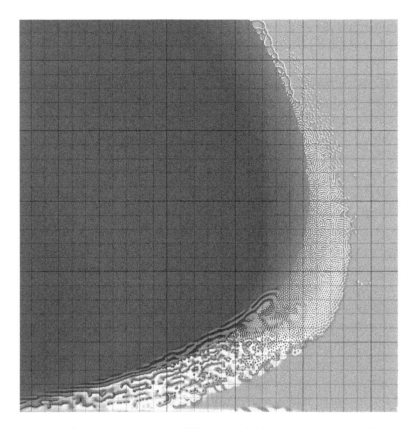

Figure 5. The Gray-Scott reaction-diffusion model for two reactants, *U* and *V*. Feed *rate (F)*, the rate at which reactant *U* is added to the environment, increases from the lower end of the figure to the upper; *removal rate (k)*, the rate at which reactant *V* is drawn out of the environment, increases from left to right. (Variation in the concentration of *U* is represented by variation in gray saturation; concentration of *V* is not represented in the diagram.) Under certain combinations of *F* and *k*, moving figures emerge. Courtesy Robert Munafo.

including tools and instruments, machines, built structure, and so on? Or are they principally enactive and, by extension, somatic? Or do we see tradeoffs, historically, between reliance now more on external technology, now more on techniques of bodily regulation? Or is this a false choice? Is even the most thing-centered strategy of survival ultimately founded on the evanescent palpable traces (gestures, words) of an enactive strategy?

This way of phrasing sounds tendentious, as if I'd set up the technological view to look bad. But this is an artifact of how dominant the technological view has become. We take it so much for granted that any effort to decenter it appears as a challenge to received wisdom.

In this chapter I have sought to clarify why this challenge is important. When we think about the changes underway in the human environment—the environment humans share with a good proportion of the Earth's living things—we tend to think in technological terms. The proximate causes of urbanization and climate change are technological, or, at least, mediated by technology in visible, easy-to-grasp ways—thus the implications must be technological too. We imagine technological consequences—shortages of water, fuel, and building materials; a failure of agricultural technology; et cetera. We imagine technological solutions in the form of a new portmanteau of material things with a smaller carbon and energy footprint. We imagine, that is, that having exhausted the resources of the basin of attraction in our adaptive landscape that has served us as a species for the past eight generations (though with great and now increasing inequality of yield and at great cost to other living things), we must simply look for a neighboring basin of attraction that is less prone to exhaustion. In fact, a better characterization of our present evolutionary moment would be to say that we have altered the adaptive landscape itself in ways that demand a reconsideration of who we are as a biological phenomenon. We need a public conversation about what the environmental crisis demands of us not as consumers of material things but as biological beings whose first and last interface to the world is the body.

4boro Landscapes and Scaffolds

This chapter represents a patch on the preceding. It overlaps thematically with the first half of chapter 4 but takes a different approach, one that, while it does not represent a substitute for the approach taken above, better highlights a key theme of the book: the complementarity of landscapes and scaffolds as guiding rubrics for making sense of how behavior changes over time. You can read it out of sequence, on its own, or not at all.

In science, as in life, metaphors shape our sense of what is possible—above all, our sense of what is, to use a word beloved of technology entrepreneurs, adjacent. So perhaps there is no more urgent task for the philosophy of science than to clarify and probe the going metaphors that underpin the scientific enterprise—to show where they come from, what assumptions they entail, whose interests they serve, and what they render invisible, and to offer alternatives that shift our sense of what is possible and what is valuable. Perhaps there is no more urgent field in which to practice this task than evolution.

Evolution is one of those things that exceed our episodic grasp. It is at once too big, too small, too slow, too fast, and too widely distributed to project on our inner screen. The conceptual devices we use to make evolution tractable in imagination shape how we respond to the confluence of adaptive challenges facing our species, not to say most living things on Earth, today. These include climate change, naturally, along with urbanization and the ongoing revolution in information technology, including technologies of biological information. There is no optimal master metaphor, no "one best way" to reason about these challenges. But by becoming more aware of how we do reason about them, and what alternatives we have, thus far, declined to take up, we enhance our odds of coming through the impending adaptive bottleneck intact.

In evolutionary science, spatial metaphors abound. Biologists, anthropologists, linguists, economists, and philosophers all speak of adaptive landscapes. By this they mean, loosely, the set of possible combinations of traits that some living thing or community of living things might possess. Different trait combinations entail different strategies for getting on in the world—perhaps you eat little meat and commute by scooter while your neighbor keeps a diet rich in animal protein and cycles. The adaptive value of a particular combination of traits—a location in the landscape—is the degree to which it tends to contribute to your flourishing. Adaptive value depends on your own history, what others are doing, and environmental factors—perhaps commuting by bike is less likely to contribute to your flourishing when it's 45 degrees Celsius out.

The landscape metaphor encourages us to imagine that the topology of evolutionary space—essentially, what it means for two points in the space to be adjacent—is similar to that of the agrarian environments we associate with the word landscape: a terrain of rolling hills, say, that we cross on foot, looking for firm ground. But is this how evolution works? In this chapter I argue that to understand the implications of changes underway in our environment—climate change, urbanization, the unprecedented mobility of capital, people, and information (including misinformation)—we need to let go of the landscape metaphor. We need to embrace a richer, more challenging picture of how strategies of survival and flourishing get enacted in our bodies and communities and how this enaction unfolds over time.

The introduction of the landscape metaphor into evolutionary theory is widely attributed to the geneticist Sewell Wright, who in a 1932 paper spoke of "fitness landscapes." The statistical theorist Ronald Fisher was working with a similar concept around the same time. But it is not difficult to imagine the analogy from trait combinations to landscape—in Wright's and Fisher's case, combinations of different alleles at different gene loci—arising independently many times. Our faculties of conceptualization are rooted in our faculties of spatial reasoning. In the neuroanatomical dimension, aspects of the posterior parietal cortex focally implicated in temporal, social, and abstract conceptual reasoning appear to have originated with the exaptation of functional networks initially selected for their value in allowing an animal to maintain a consistent postural, kinematic, and perceptual relationship to objects in the nonself world. This is just one of many ways in which, over evolutionary time and over developmental time (childhood, adolescence), *movement precedes thinking*. Indeed, it is not just that movement precedes thinking, but that thinking represents an internalization of movement. It is not that we have bodies to give expression to a will that originates in the nervous system. Rather, we have nervous systems to afford the body responsiveness to change in the environment in the frequency bands characteristic of motile as opposed to sessile living things.

Developmentally and culturally, our experience of space shapes how we organize the world conceptually. We see this, for instance, in how we organize language. The linguists Stephen Levinson and Niclas Burenhult describe what they call *semantic templates*, devices of conceptual organization that recur across the lexicon, drawing together words from different domains of interaction (home, work) and different word classes (nouns, verbs) into conceptual sets that occupy distinct relationships to one another. Often—at least, in the instances most salient to outside observers—the relationships that give form to the semantic templates operative in a community are derived from forms of spatial experience salient in that community. So, for instance, in hilly places, you might observe that words appear to be organized into *uphill*, *downhill*, and *transverse* (imagine crossing a hill along a fixed isopleth) sets. In places dominated by tides, you might see a template based on the experience of going *with the tide* or *against the tide*. Semantic templates represent environmentally specific outgrowths of a broader phenomenon in how we make sense of the world, what philosopher Mark Johnson and linguist George Lakoff have called image schemata. Some image schemata appear to be universal—think of smooth–rough. A smooth–rough distinction is salient to infants, who associate smooth- and rough-surfaced objects across the senses. There is something that it looks like, we might say, for an object to feel a certain way. This may sound like synesthesia. But, in fact, a significant part of our sensory experience is metamodal—that is, it refers to qualities in the complex patterns of recurrence, across time and space, of basic units of sensory experience that are palpable in different senses and comparable between them. In other cases, cross-modal correspondences between different kinds of sensory experience appear to be rooted in the qualities of attention they induce—so, for instance, acrid odors and plangent sounds feel "sharp." Some image schemata, for instance inside–outside, are topological.

So our habits of conceptualization are rooted in *body schema*, that is, our strategies—as individuals and as members of communities—for negotiating the interface between body and world. We are practically compelled to organize the world topographically. Indeed, spatial reasoning represents so basic a resource for how we reason about other things that it is nearly invisible. We speak of concepts as being neighboring to or distant from one another. This tendency is helpful up to a point. Where (again, a spatial metaphor) it becomes unhelpful is when we try to specify the topology of some abstract space (say, "the space of everyday behavior"—recall the example of diet and commuting strategies from the first paragraph) and find that that topology diverges from the topology of space as we experience it with our bodies.

Let's think for a minute about what it might mean for a topology to diverge from that of space "as we experience it with our bodies." Earlier, I characterized adaptive landscapes, loosely, in terms of *the set of possible combinations of traits that some living thing or community of living things might possess*. This is a start. It allows us to imagine a number of different dimensions of variation in behavior—returning to our "everyday life" example, we might include diet, commute, perhaps what you like to read or watch, whether or not you have a partner, and so on—and to organize combinations of different values along these dimensions as ordered tuples. An ordered tuple is simply an extension of the ordered pairs familiar from basic geometry. We might reserve one dimension—one slot in our ordered tuple—for adaptive value or "the degree to which this strategy contributes to flourishing."

But this definition is missing something essential: a sense of how tuples with different values in different slots fit together to form a fabric, a space. In other words, we need to understand what makes one tuple a neighbor to another. One way to do this is to specify the neighborhood for each tuple by exhaustive enumeration: in other words, we say, "For the tuple with values such and such, the following tuples are neighbors," and we do this for all the perhaps infinitely many tuples in our space of strategies. Sometimes, this is the best you can do. But ideally, rather than enumerating the neighborhood for each location in our landscape, we'd elicit consistent, general patterns of distance valid across the landscape. We'd like our distance relation to be roughly transitive, so that if *a* is near *b* and *b* is near *c*, we know something about the distance between *a* and *c*. Perhaps this "something" is simply, "*a* is no more than two hops from *c*"—that is, distance is something we measure by *path segment cardinality*, the total number of segments in a path, much the way you'd count up the number of seats in a train compartment to measure its capacity without considering the fact that bodies vary in size.

But really—and this is where we run into trouble, for we're making assumptions based on the way physical space works—we'd like our distance relation to be a *metric*: that is, we'd like to be able to define uniform cardinalities or magnitudes of distance between any two points in our landscape and be able to add distances up and scale them by a constant factor.

Even now, we have not yet constructed a metaphorical landscape sufficiently like what we encounter out in the world to safely pull in all our assumptions about "how landscapes work." For one thing, we have not yet specified our metric. Maybe this seems needlessly pedantic: Is there more than one way distance can work? How could distance be other than Euclidean distance, that is, the square root of the sum of the squares of distances in each of the dimensions of our ordered tuples? We might need to make allowances for the fact that Euclidean distance fails when the land is not perfectly flat—but surely it can serve as a useful first approximation?

Or perhaps not. Imagine you are standing on a corner in an urban environment with streets and lots laid out on a grid. This is a relatively mature city: every lot is either occupied by a building or is fenced off for construction. Your only option for getting from point A to point B is to follow the street grid. Imagine the streets are placed equidistant on the grid. That is, every lot is a square, and "one block" means exactly the same distance no matter which way you're walking. You're new in town. Say point A is where you're staying and point B is a coffee bar where you've agreed to meet a friend. If a passerby tells you point B is "three blocks up and one over," the distance you need to walk, L1, is $3 + 1 = 4$ blocks

Now imagine the buildings and fences have been erased. You see your destination off in the distance and strike out in a straight line for it. Now your distance, L2, is $\sqrt{(3^2 + 1^2)} \approx 3.17$. The character of the landscape—the conditions of adjacency, of neighborness, that it affords—determines the metric. The L1 or "Manhattan distance" metric generally does not coincide with the more familiar L2 or Euclidean metric, save in cases where the path between two points lies along one of the basis dimensions (the axes of the street grid) in the L1 world.

L1 and L2 are simply two kinds of distance that we encounter commonly. They do not begin to exhaust the ways an adaptive landscape might be organized topologically (or, as in the case of a simple enumeration of neighborhoods, *pre*topologically). And the fact that we cannot know, in advance, what the topological structure of an adaptive space looks like is just the start of our problems. For adaptation unfolds in a number of different media simultaneously, and the topology of the adaptive landscape in one medium may be at odds with that in another medium.

By medium here I have in mind things like the genome, epigenetic modulation of gene expression, transient physiological states, and patterns of recurring behavior such as those I alluded to at the top. You might be inclined to think of these as different "layers" of the evolutionary process—but "layers" carries a sense of hierarchy (again, an image schema—*layers progress bottom to top*), with the lower layers in some difficult-to-make-precise way "more fundamental" than the upper layers. This is how many of us tend to think about the relationship, say, between genes and behavior, or the physiology of the brain and cognition—the latter is built upon the former and in some way caused by it. When we imagine the world as a layer cake, we naturally tend to assume that the upper layers are less important for understanding the ground truth—the bottom layer—of behavior. But to take a layer-cake metaphor for granted is to prejudice our understanding of the relationship, say, between genes and behavior. So we need something more neutral, something that helps us be mindful of the fact that causal arrows point in all directions—from everyday behavior to epigenetic modulation of the genome, for instance, or from emotional states as we experience them to brain states. One way to do this is to think of the adaptive process as a network, with the same process unfolding in parallel at all the nodes in the network and with lines of influence—"edges" or "arcs" in the network—extending between the nodes. The different nodes have different media, and what it means to "walk" from one location in the adaptive landscape to another means something different in each of these media—the landscape at each node has its own topology, independent of those of other nodes. The process of moving through the landscape unfolds simultaneously across the whole network—each node is one representation of a single phenomenon of adaptation, but each representation carries slightly different information about that process. And the correspondences between these different representations are not one-to-one (or, to use more precise language, isomorphic, "of the same shape").

One way to grasp what the adaptive process "looks like" in the network of media is to imagine you are standing in a gallery watching a video installation. There are three screens before you, each with a scene unfolding. It is the same scene, and the three films are timelocked. But each was filmed from a different point of view, so that what is salient in one version may be invisible in another. Another way is to imagine you are back in the unfamiliar city, trying to get from home to coffee bar. But this time, your consciousness is divided among four possible worlds, so that you experience the walk from A to B simultaneously in each of them. In one of them, the space between starting point and destination is laid out on a rectilinear grid of streets. In another, the environment is more like that of London or Tokyo than New York—a concentric series of spoked hubs, so that perceived distance is largely a matter of the total number of segments in your path, but not in the same way as in the L1 space. In a third, the space between you and your destination is an open field, so your best path is a straight line. The fourth world is pretopological—origin and destination are arbitrarily adjacent, so the perceived distance is, say, "one unit of time" or however long it takes to move from a point to one of its neighbors. In all four worlds, the journey you make is the same in that it has the same origin and the same destination. But the experience of making the journey is different in each world. In some of these worlds—Manhattan, London/Tokyo—the journey carries more information than in others. As far as you are concerned, all four journeys happen at the same time, super-imposed one atop the other.

At this point, we must step back from the metaphysical cliff edge. (Indeed, there is a branch of mathematics, category theory, dedicated to the characterization of what I have called "networks of media." Category theorists refer to their work, approvingly, as "abstract nonsense.") My point is simply this: image schemata and spatial metaphors, for all that they facilitate our reasoning about phenomena that are too big, too small, too slow, too fast, or too widely distributed to project on our inner screen, can lead us astray. Such is the case with evolution.

The multiplicity of the topology of evolution is one challenge we must address if we are to repair the landscape metaphor. Another is the fact that "search" or "diffusion"—moving from one location to another—is not always the best way of understanding how the evolutionary process unfolds. Consider the genome. One of the major drivers of the formation of new evolutionary clades over geological time is *polyploidization* or whole-genome duplication. Genome duplication expands the "space" over which evolutionary search unfolds in subsequent generations in the lineage that has experienced the duplication event. Broadly similar phenomena arise in other media of evolution—new body-plan elements (limbs, say), new sensorimotor faculties, new technologies (for instance, controlled use of fire) that create or open up previously inaccessible regions of adaptive space. Sometimes, the topology of a medium changes, or the character of the diffusion process unfolding in that medium changes, as with the emergence of cell-to-cell signaling machinery in the early stages of multicellularity. If this were not enough, we must also include nodes in our network for various dimensions of change in the environment—that is, the evolutionary process is one of niche construction, with selective cues flowing in both directions between a community of living things and its environment. And, as I noted in the previous chapter, *we must take into account the fact that often we observe multiple communities of living things with nominally*

unrelated germlines evolving as metacladal unities or holobionts—
think of the relationship between humans and the microbiota that
colonize our orifices, skin, nervous system, and gut.

As these last comments suggest, there are in fact two kinds of
edges or "lines of influence" between the different domains or media
of adaptation. Above we had in mind the structural edges—abiding
relationships between the topology of one medium and that of
another. But evolution is a process that unfolds over time, and the
edges that interest us most tend to be the dynamic ones, those that
describe how change in one medium conditions change in another.
Imagine these dynamic edges as series of bright dots flowing from
one domain to another—or, just as often, looping from a domain
back to itself. Each dot represents a single instance in which events
in the source domain condition events in the target domain (in category theory, the "co-domain"). To take an everyday example, consider
what happens when you have your first cup of coffee, tea, or whatever your stimulant alkaloid of choice, in the morning. About twenty
minutes after you take that first sip, you start to become aware that
the world has changed. You experience a new alertness and capacity
for concentration, perhaps a new expansiveness of mood, flights of
imagination, analgesia if you've been in pain, suppression of appetite
if you've been hungry—or renewed appetite if you've been inclined
to inanition. To the extent that this experience is pleasurable, and
to the extent that you associate the experience with the sip of coffee,
say, you become more favorably inclined to the taste of coffee. Most
conditioning phenomena unfold in drip-like fashion, by degrees,
over time. The individual drips are the selective cues, the signals that
nudge us, as communities and individuals, toward new forms of life.

The looping of selective cues from a domain to itself is ubiquitous, but in the domain of behavior it takes on a special quality, that
of self-awareness or reflexivity. Not only does behavior direct cues
to the epigenetic, genetic, and environmental nodes in our network
of corresponding landscapes, it also, as it were, interferes in its own
evolution. For as soon as a pattern of behavior becomes distinct

enough or common enough—as soon as it crosses the threshold of social salience—it becomes an object of commentary and an index of a host of superficially unrelated commitments: Are you a meat-eater or a vegetarian? Do you cycle to work or drive? In humans, the social-indexical processes that take form around the regimentation of behavior are themselves a big part of behavior. (Think of language: think of how much of everyday speech is concerned not with "conveying information" but with calibrating modes of address, signaling affinities, maintaining or challenging the status hierarchies coded into how we speak to people with whom we have different kinds of relationships.)

Topological heterogeneity, causal multiplexity—these are among the challenges in understanding the relationship, say, between how a change in environment shapes our behavior and how it shapes the epigenetic modulation of the neuroendocrine signals that mediate stress response. Adaptive landscapes are a useful conceptual device in making sense of relationships of this kind. But they are not enough. We need more than one metaphor. We need a whole toolkit of conceptual abstractions if we are to formulate a language for talking about evolution, above all the evolution of culturally patterned behavior, that is both rich enough to capture the complexity of the evolutionary process and distilled enough to support imaginative thinking about possible futures. What might such new conceptual tools look like? One that I have found generative is the scaffold.

A scaffold is anything that both supports and constrains the development of something else. Think not of the structures erected to facilitate the repair of building façades but of those used in tissue culturing, as in the case of the fabrication of skin grafts or, in proof-of-concept form, cultured meat. For that matter, think of the scaffolds that occur in living things all around us: the xylem that gives structure to the ligneous stems of trees, the fibroin that gives silk its distinctive tensile properties. In tissue fabrication, a scaffold may be something as simple as a sheet of absorbent paper imbued with an extracellular growth medium to convey nutrients to the cells under culture. As a conceptual device, the value of scaffolds is that they turn our attention away from the process of searching a strategic landscape for some stable, highly adaptive spot and toward the ongoing relationship between media of evolution that constitutes the adaptive search in action. Above, I characterized this relationship as one of *conditioning* or a flow of selective cues. But the image I offered—a train of dots— suggested a rather arid and distant relationship. By contrast, when we imagine these conditioning relationships as forms of scaffolding we understand them as intimate and enduring processes—sometimes, the scaffold becomes embedded in the new thing, be it a sheet of muscle or a pattern of behavior, that has grown up on it.

Scaffolds offer at least two further advantages. The first is that they simplify the task of holding in our minds the ongoing crosstalk between different media of evolution. We are no longer obliged to imagine corresponding landscapes with distinct topologies. Instead we are led to focus on the dynamic character of that correspondence. The third selling point for scaffolds is that they help us push back against the "view from nowhere," to adopt philosopher Thomas Nagel's phrase, that tends to accompany the landscape view. If you are anything like me, when you imagine adaptive search through the lens of the landscape metaphor, "you" are somewhere off in the distance, perhaps standing on a rise, watching some other being move through the landscape looking for a stable place to abide (a "basin of attraction"). The tendency to adopt a view from nowhere is exacerbated when we think not of search but of diffusion—a community spreading out across the landscape, so that different individuals adopt different strategies but the ensemble, the way these strategies function together, is itself stable and adaptive. The view from nowhere has its role, but, again, it has its limits too. Among these limits is that it masks the degree to which behavior itself, our own behavior, serves as a scaffold.

We are accustomed to thinking of the guiding hand in adaptation, that which directs us as we move about the adaptive landscape, as something "outside us"—preferably, outside the domain of living things. Consider technology. We think of technology as something that gets realized in material artifacts. We incorporate these artifacts into our lives and in this way our bodies change—habits of posture and movement; patterns of skeletomuscular development and joint wear; qualities of attunement to movement, light, pressure, and sound; circadian and longer-term rhythms of motor vigilance, mood, and metabolic activity—the entire constellation of phenomena that make up our somatic niche. Technology serves as a scaffold for somatic niche construction.

But often, the guiding hand—the scaffold—consists not in technology in the conventional sense, that is, a repertoire of material artifacts and the behaviors associated with them, but in enactive artifacts, shared patterns of behavior that exist by virtue of ongoing bodily enaction rather than by realization in material things. One way to read the history of technology is as a history of the trading off of survival strategies centered on material artifacts for those centered on enactive artifacts. But since enactive artifacts do not fossilize, past instances of societies' prioritizing enactive artifacts over material ones are often not salient as instances of innovation—to the contrary, they often come across as instances of maladaptive technology loss—as we saw in the case of Tasmania. This makes it difficult to imagine strategies, say, for adapting to climate change that do not entail simply substituting a new package of material artifacts with a smaller carbon footprint for those we have today.

Earlier, I described adaptive landscapes as partial representations, partial views, of the adaptive process. Each view makes salient certain features, certain dimensions, of the whole. In mathematical terms we might call the landscapes *manifolds*. I won't bore you with speculation on how we might characterize scaffolds mathematically. I simply want to stress that scaffolds and manifolds offer complementary ways of approaching the same phenomenon.

To return to our earlier example, rather than speaking of the recurring experience of coffee's effect on mood and alertness as a train of cues that select for a taste for coffee, we might now speak of coffee

Figure 6. Two ways of imagining evolution. In the *network of milieux* view, each node in the network is equipped with a distinct topology and each serves as partial substrate for a process of *search* or *diffusion* that unfolds across the network, with influence between the nodes. In the *scaffold* view, certain recurring patterns of behavior serve both to support and constrain the development of more elaborate patterns of behavior. Illustration by Yoshi Sodeoka.

as a scaffold for a quality of mood and alertness that, for whatever reason—perhaps it is simply pleasurable, perhaps it makes work feel less tedious—contributes to your flourishing. But it would be more precise to say that the scaffold in this case is the behavioral complex *coffee-drinking* rather than the substance coffee. That is, we would do well to attend to how ritually demarcated patterns of behavior serve as scaffolds for our adaptation—physiological, epigenetic, behavioral, et cetera—to conditions that, if we take the long view, are themselves largely of our own making. In the case of coffee-drinking, these conditions might include the pervasive use of artificial light and, more recently, the shift to a cognitive form of capitalism that prioritizes distinctive qualities of alertness and concentration. Coffee-drinking has become an element of our somatic niche.

Scaffolds, as I've outlined them, do not represent a universal solvent any more than landscapes do. But they do offer a new perspective. The next time you read a news item about what urbanization, or climate change, or social media is doing "to our brains," consider these phenomena in two ways. First, imagine them as processes of diffusion through an adaptive landscape—What is guiding us into certain basins of attraction, and what consequences could we imagine if we remained there? Second, imagine them as processes of behavioral scaffolding. How does urban living, say, or the increasing incidence of temperature and humidity extremes, condition our behavior, including our characteristic ways of using our bodies? Rather than imagining strategies for modulating these effects as a search for some better place (strategic or geographic), ask what kind of concerted, reflexive, coordinated changes in bodily behavior might allow us to reshape ourselves, and our environments, for a long future.

5 Ditch Kit

This morning I am in Los Angeles. It is the second week of December. The summer and autumn were exceptionally dry across California, and the fire season rivaled that of previous year, alluded to in chapter 4. A month ago, in the foothills of the Sierra Nevada some 750 kilometers to the north, the Camp Fire destroyed the town of Paradise. But in the past couple weeks we have had rain, and this morning it is overcast, about 17 degrees Celsius. As in other winter-wet climates—Southwest Australia, for one—the houses here are not built with the cool-wet season in mind. Often it is colder inside than out, and the baseboard heater is of little use in the broad front room of the cottage where I've been living unless you are standing directly in front of it, which is where I am standing now. When I step outside, I catch the peppery-sweet scent of gum oil, giving me the uncanny sensation that I am back in Australia. Gums are not native to this place, but they thrive in this climate. On the fire trail at the top of the ridge behind my house, ghost gums loom up among the red pines and cypresses, the smooth, cream-colored bark and slender arborescent habit suggesting they are *Corymbia aparrerinja*, though I would not trust myself to make an identification. Eight years ago, when I first moved to Berlin, I damaged the vasculature of my toes, and now they get chilblained even in mild cool weather such as you get here. I find myself missing winter.

A year ago, when I first started thinking about this book, I wrote the following. It came in a single, fluid stroke—it could not have taken me more than half an hour to write it. For the life of me, I cannot now come up with a better way to express my intent with this book:

> *Ditch Kit* is a meditation on the process of sloughing off material things, of reducing the stock of one's worldly possessions to something that fits in a knapsack. It is about becoming intimately familiar with the practical significance of volumes and stowage spaces: what fits in twenty liters, in thirty, what fits under an aisle seat as opposed to a window. It is about becoming obsessed with paring down one's wardrobe, cutting away redundancies and indulgences, formulating a uniform. It is about the odd blend of precariousness and privilege that comes with living this way and with how renouncing consumer excess becomes a form of consumer excess in itself as the effort to identify the one object that will take the place of ten takes over one's life. It is about the substitution of corporeal technologies—of exercise, of thermoregulation—for material ones. Above all it is about stillness, about the longing for a place to unpack and hear silence, or at least something other than the spectrally unstructured hum of transit spaces.

Having explained what I wanted to do in terms that made sense to me, I had then to explain it in terms that would make sense to patrons and editors, who would want to see a sense of continuity with what I had done in the past. I wrote this:

> *Ditch Kit* represents a successor to *The Meat Question*. It draws together a number of strands in my thinking over the past five years. One of these concerns the deep history of technology and the degree to which different kinds of technologies are archaeologically legible and thus salient for our sense of where we come from, where we are going, and the mode and tempo of cultural evolution. A second concerns the human enmeshment in the world of living things and the

nature of our relationships to other large-bodied animals, to trees and other plant life, and to the microbiota that form the greater part, by information content, of the holobionts that we are. A third concerns social acceleration and its concomitants: climate change, economic volatility, and growing interest in bodily techniques for fostering mindfulness and resilience that often serve to paper over rather than address the injustices of social acceleration. Above all, *Ditch Kit* aspires to cast light on the violence of the ascetic impulse, the ways in which renouncing objects and even a fixed living situation in the interest of achieving stillness entails complicity in the accelerationist churn that makes stillness, silence, and material security increasingly inaccessible to most people in the world.

In between these paragraphs came a third, less relevant for our purposes save that it will help you make sense of my cryptic remarks about *the spectrally unstructured hum of transit spaces*:

Over the past ten years, two things have dominated my encounters with the built world. One has been transitory living and an increasingly vexed relationship with objects. The other has been tinnitus and an increasingly vexed relationship with sound. I hear monaurally, in just one ear, so I can neither localize sounds in space nor extract speech and other auditory objects from noisy backgrounds. As tinnitus and related pathologies of hearing have become standing features of my acoustic experience, I have started to see how deeply monaurality has shaped me as a body and as a person. A couple years ago I began making sound recordings with the aim of making sense of how sound participates in our sensorimotor, affective, and political lives, above all our encounters with pain. Lately I have come to see that movement is equally implicated in pain, that the punctuated stasis of transitory life, the periods of bodily confinement and waiting, stand in an antagonistic relationship with stillness in both its acoustic and motoric senses.

The relationship between sound and pain will have to wait for another book. What of the rest—*the violence of the ascetic impulse, the substitution of corporeal technologies for material ones, the human enmeshment in the world of living things*? Over the past year I have struggled with these themes, convinced that they represent aspects of a unitary phenomenon. In Tokyo in October 2017, in a guesthouse in the western suburb of Hibarigaoka favored, for its reasonable rates and relaxed atmosphere, by Cameroonian migrants, I sat on the bed in a cold room. Jessy had left for Switzerland to thread another bead in her own itinerant career. Our time in Shiga was receding, though outside another late-season supertyphoon was dumping rain on the laneways and convenience stores of suburban Tokyo. I had been asked to write something for an upcoming event in Berlin. The theme was "the technosphere." More specifically, I had been asked to offer a long view of the role of randomness in technology. My text was to be performed at an event that would also include a reading from *A Million Random Digits with 100,000 Normal Deviates,* a book, as its title suggests, of random numbers, produced by the RAND Corporation and published in 1955 as an aid to the implementation of Monte Carlo simulations in the days before robustly pseudorandom permutation algorithms, and the hardware to implement them efficiently, were widely available.

I wrote about fire—parts of this text appear in chapter 3. At one point, sitting cross-legged on the sagging bed in Hibarigaoka, dressed in two pullovers and a watch cap, the rain slanting down on the parking lot outside, *The Road* still fresh in my thoughts, I wrote this: "The world is ending. What do you take with you? What is in your bugout bag, your ditch kit? What if your ditch kit had to fit in an Altoids tin? Would it include a firesteel and ceramic striker?" The *ditch kit* felt like the unifying phenomenon, the *gluon,* to use philosopher Graham Priest's term, the property that participates in all the constituent parts of some complex thing and in this way binds them together.

One day four years ago—November 2014—I went to see an apartment. This was in Berlin. I needed a place for the winter at least. The sky was the unrelenting gray you see there from October through March, and in my memory it was cold. It was early in the season, but already I could feel the tension between the shoulder blades that comes from trying to make your body as compact, as thermally efficient, as possible. Five flights up, I was met by the graphic novelist S. "Come in," he said. "Nothing in this house is mine." This felt to me like a distillation of my life the previous six years.

S made tea and explained that he did not hold the lease, he was simply a subtenant. The lease was held by a third man, whom I would meet later. S would be away for the winter. From the bedroom window, if you held your body so that your line of sight ran nearly parallel to the wall, you could just see Maybachufer, where the Turkish market was held Tuesdays and Fridays. I told him I would take it, my thoughts turning to the task of packing up what worldly possessions I had and getting them here.

Today, of course, anxiety about stuff is widespread. We brood about the role stuff plays in our lives and the hold it has over us. In this instance, when I write *we* I do not have in mind people with legitimate cause for concern about the disposition of their home and its contents (clothes, cooking equipment, mobile phones)— refugees from civil war in Syria, state and communal persecution in Myanmar, hyperinflation in Venezuela, or urban warlordism in Central America, not to say the growing number of people in Los Angeles and other major cities in the western United States who carry on with working life during the day but sleep in their cars at night because they cannot afford housing. The displaced and the unhoused represent large constituencies—as of the end of 2017, according to UN High Commissioner for Refugees figures, the number of people in all parts of the world who were "forcibly displaced," whether within their country of origin or internationally, came to 68.5 million. But it is not the displaced and the unhoused who have made Marie Kondo's *The Life-Changing Magic of Tidying Up* a bestseller.

Kondo's manifesto appeared in English in October 2014, about a month before I found myself in S's kitchen on Hobrechtstraße looking for a place to finish *Computable Bodies* and wait out the winter. At the time, I was considered a bit eccentric for the facts that the sum of my material possessions fit in a pair of knapsacks and that I seemed to care more about the design and construction of the knapsacks than about what was in them. In the years since, this way of living has become, if not common, then a common aspirational template. Decluttering and its edgy younger sibling minimalism have become behavioral emblems of a new genre of relationship to the ensemble of material things that sustain us through the day.

We are inundated with instructional material. In addition to Kondo's own copious output, the past four years have seen a number of new entrants to the decluttering category. As in previous reconfigurations of our relationship to the material world, much of the energy behind the decluttering movement emanates from Japan and Scandinavia. There is Fumio Sasaki's *Goodbye, Things*. There is Shoukei Matsumoto's *A Monk's Guide to a Clean House and Mind*. There is Margareta Magnusson's *The Gentle Art of Swedish Death Cleaning*.

The recent upwelling of interest among, let us say, educated professionals, in reducing their material footprint is driven by at least two things. One is diffuse if heartfelt concern about what you might call our crisis of stuff—the ballooning quantity of microplastics in the oceans and in our bodies, the aerosol pollution alluded to in chapter 4, some of it originating with the artisanal disassembly of consumer electronics in periurban places in Guangzhou and Ghana. The other is that phenomenon I alluded to above as precariousness. Precariousness is, among other things, a state of suspension, an undefined wait similar to the undefined waits we experience in airports or hospitals—*undefined* in the sense that we have no clear sense of how long the wait will last. It is these everyday undefined waits that I had in mind a year ago when I wrote, in the notes that became the seed for this book, of the *punctuated stasis of transitory life*. For more and more people, life history, the staged unfolding of behavior over time, is characterized by a similar quality of punctuated stasis, with extended periods in which one lacks the means to do more than subsist—perhaps as a digital nomad, or in one of a proliferating number of "dormitories for grown-ups"—punctuated by episodes in which what entitlements one does possess—to housing, professional identity, and so on—rapidly come undone. No measure of education, professional socialization, or effort feels sufficient to guarantee the continuity in these entitlements necessary to imagine the arc of one's life as a unity.

In chapters 2 and 3 I proposed three horizons of niche construction—the episodic, the biomic, and the somatic. I formulated these horizons with reference to an account of how humans had shaped the biosphere, and how it had shaped them in turn, in two distinct late-Holocene biomes: the cool-temperate maritime woodland of Tasmania and the warm-winterwet open woodland and grassland of Southwest Australia. These, of course, do not exhaust the human adaptive range, which spans the gamut of Holocene biome types from polar tundra to tropical summerwet. But they were enough to allow us to see that the human engagement with the nonself world is distinguished by the degree to which it is reflexively elaborated— that is, the degree to which we are aware of what we are doing—over time horizons that exceed not just the episodic, the here-and-now, but the lifespan of the individual. Our case studies in chapters 2 and 3 also served to demonstrate how difficult it is to reason about planning. Planning is marked by an indeterminacy that is both epistemic and ontological. Often we cannot know which horizon of effect was foremost in the minds of the individuals or community who enacted some transformative process on the biosphere—burning, say, or domesticating plants and animals, or establishing dense, permanent settlements. Our inability to know is not simply for want of evidence. It is for want of clear demarcations in the temporal extent of the will to modify the environment. Planning is an emergent phenomenon.

In chapter 4 I argued that this way of breaking out the mode and tempo of human activity, and of reasoning about its conceptual indeterminacy, is useful not just for thinking about the lives of Pleistocene-to-late-Holocene foragers. It applies to our own lives, and our own environmental challenges, too.

But whatever the challenges of identifying and reasoning about planning, we can feel secure in saying that precariousness represents a contraction in the horizon of planning—as this horizon intersects the arc of one's own life and, perhaps, as it intersects the arc of our species' time on Earth. This is a theme I return to below.

The crisis of stuff, precariousness: these are two ways of reading the appeal of decluttering and minimalism. A third is capitalism. Again, from my notes:

> Sometime in the last couple years, perhaps around when *kondo* became a verb in English, minimalism seems to have lost its way. What began as a movement dedicated to exposing how having more than we need makes us sick, engenders in us an insatiable sense of want, has become another medium of consumer aspiration. Or that is one way of reading the situation. Another is that decluttering has always been about shoring up the capitalist imperative to accumulate, that "tidying up" represents a way of rationalizing and containing the unruly purchase stuff has on our lives.

The fantasy of sloughing off the weight of stuff did not originate with capitalism. Basho, the Edo-era poet credited with formulating the haiku as we know it today, alluded to the depression he experienced when a disciple arranged a house for him on the southern fringe of what is now Tokyo. This was 1680. He wrote, that autumn and winter, of lying awake at night, "the sound of an oar hitting waves / freez[ing] my bowels" and of sitting at home, alone of a snowy morning, gnawing on dried salmon. Later, when he took to the road in the first of what became a series of long walks, he described a new feeling of lightness: "As I have no home, I have no need of pots and pans. As I have nothing to steal, I have nothing to fear on the roads."

Not having a home, of course, does not entail having no need of pots and pans. It is simply that the pots and pans one has need of are no longer one's own. Uncoupling the use you get from implements of everyday sustenance from a stable possessory relation of the sort we call ownership is one way to manage the unruly purchase of stuff. This is an approach with deep roots and growing appeal. There is something at once courageous and humbling, at least in fantasy, about renouncing the comforts of home and setting off, exposing oneself, as it were, to the good or ill will of strangers. From another perspective, renouncing ownership represents a form of emotional freeriding. Someone else must remain invested in pots and pans for you to free yourself of mundane entanglements. For Basho, the story of how Śakyamuni, the historical Buddha, left the palace to which he was heir to practice austerities in the wilderness served as a template of ethical conduct. But there is nothing specifically Buddhist about this fantasy, or rather, what is buddhist—small *b*—about it comes not from its historical relationship to Buddhism but from a tension that is immanent in settled life.

Just as certainly, while the fantasy of sloughing off stuff did not start with capitalism, there is something about the way we live today that makes the need to rationalize and contain our relationship to stuff particularly urgent. There are, we learn as we grow up, appropriate and inappropriate ways of relating to stuff. Keeping a tidy home is appropriate. Hoarding and fetishism are inappropriate—save, perhaps, when they occur under the sign of capital accumulation. Having too little stuff may also seem, to some observers, in some contexts, inappropriate—a sign of willful immaturity, a refusal to commit to relationships of the kinds that depend on having a kitchen, say, or a car. If too much stuff makes one sick, too little stuff leaves one etiolate, epiphytic, undeveloped as a person.

BORO: GAMES WITH STUFF

One concern readers of early drafts of this chapter had was that the discussion here, in contrast to that earlier in the book, is too narrow in scope, too culture-bound, too class-bound. So by way of a boro on the preceding, let's consider an alternate formulation of the phenomenon by which stuff becomes implicated in personhood. My stand-in text here is a lecture *cum* essay by the anthropologist Annette Weiner, given at the 1993 meeting of the American Ethnological Society in

Santa Fe, New Mexico. That year the theme of the meeting was "Arts and Goods: Possession, Commoditization, Representation." Weiner was an anthropologist of the Western Pacific, a part of the world renowned, in anthropological theory, for its role as a source of inspiration in matters of how stuff comes into the world and how people use the tactical distribution and retention of stuff to enhance their status and, by extension, their personhood, their immanence, if you will. Weiner herself had conducted fieldwork among the Massim of the Trobriand Islands—famously, where Bronisław Malinowski (recall the preface) cut his teeth as an ethnographer. Among the key tropes, the key recurring formulae, to emerge from Malinowski's work, recorded in his *Argonauts of the Western Pacific* (1922), is that of the *kula ring*, a phenomenon by which body ornaments made of gastropod shells circulate through the archipelago—necklaces in one direction, armbands in the other—in a transgenerational game of ritualized exchange by which men seek to enhance their prestige. "In kula," Weiner notes, "the most symbolically dense shells are known as 'chiefly shells.' These shells are large, aesthetically desirable, and have long prestigious genealogies of their previous owners. Some of the highest-ranking shells are kept by one player for 10, 20, or even 30 years. Given the relatively few valuables in any of these Massim societies and the unending exchange obligations, several decades is a very long time to keep a possession out of exchange. Within kula, these most prized possessions remain inalienable because their ownership is reserved for a kula elite." At the same time, chiefly shells offer a beacon of hope to less exalted players, who see in their occasional circulation the possibility that they, too, might someday ascend into the ranks of the elite.

Shell ornaments are not the only goods enlisted in what we could call *games with stuff* among the Massim. Men wield raw yams as instruments of political power—yams "are used to challenge adversaries or authenticate status, are buried with important chiefs, are left to rot to show a chief's power, and are given to women by their brothers"—while women trade bundles of banana leaves with no intrinsic aesthetic appeal or function apart from their role as tokens in an exchange game. In Samoa, finely woven mats serve as tokens in a similar game, this one linked to the use of marriages to establish clan alliances. Often, games with stuff unfold alongside and serve to lubricate the exchange of more mundane goods.

Observing these games with stuff, Weiner proposed that objects become imbued with *symbolic density*—some objects come to be "so dense with cultural meaning and value that others have difficulty prying these treasures away from their owners." These objects "circulate exceedingly slowly in comparison to less dense ones, which can be exchanged, sold, or traded with impunity."

All objects, in fact, occupy places on a continuum of symbolic density, with inalienable possessions at one end, commodities at the other. "Certain objects," Weiner notes, such as chiefly shells in the kula game, "acquire a charisma that lasts beyond one person's ownership." As a rule, this charisma is no accident—players are forever working to nudge their tokens toward the dense end of the continuum by tactically withholding them from circulation, perhaps substituting some other thing in its place, so that an object's withdrawal from the game, reaffirmed and made known by its periodic fleeting display, becomes a property not just of the possessor but of the thing itself. Imagine an osmotic potential by which the status that accrues to the individual by virtue of their skillful participation in the game diffuses into the withheld object itself every time the individual exhibits it in the hand or on the skin. Weiner sees parallels to the games she describes in the way works of art circulate in the auction markets of the metropolitan North Atlantic—today I would add the metropolitan Pacific Rim. She emphasizes that there is nothing about, say, capitalist societies that distinguishes them categorically from the subsistence societies of the Western Pacific. Everywhere, we observe prestige playing a role in the calculus of exchange. Similarly, I would add, everywhere we see people engaged in multiple bookkeeping, reserving certain caches of value for certain kinds of transactions even when, in principle, all are to be conducted in the same kind of money. Some money, say that received from a high-status patron or in exchange for a higher-status form of labor, is simply better fit for some purposes, and different kinds of money carry different obligations of redistribution among kin. People go to great lengths to formulate and maintain differentiated registers of exchange even when, superficially, the same currency functions in all of them. Events where capital in one register, be it marked by a distinctive kind of currency token or simply by a system of book-keeping, gets transferred into a more prestigious register tend to be ritually circumscribed.

In children's games we catch glimpses of how games with stuff take form spontaneously and become self-sustaining, along with how things valued, initially, for their sensuous and relational properties come to take on the role of currencies and how register boundaries emerge in the use of these currencies even in the absence of outward distinguishing marks. Consider this passage from Gerald Murnane's novel *Tamarisk Row*, which, recall, I quoted at the close of chapter 1 for its exemplary use of scale-shifting as a narrative device. The setting is a country town on the inland plains of Victoria, Australia, 1948. The protagonist, nine-year-old Clement Killeaton, intimidated by his peers at school, spends most of his free time alone in his backyard, enacting elaborate fantasies centered on the horse racing that has been the cause of so much of his father's disappointment. At one stage in his fantasizing, colored marbles come to stand in for thoroughbreds. Soon, he is moved to compile a studbook of his marbles. Observe the similarities between the antipodean chain of marble exchanges Clement imagines and the kula ring described above.

> He begins as a small boy in grade two making entries of a few words in each of a few columns. Before he has finished writing about half the marbles in his collection he gets from his father a much bigger ledger and starts a new system of entries with one whole page for each marble. . . . Towards the end of his grade-three year he reviews what he has done and counts his marbles and calculates that it may still take him several more years and perhaps two or three more volumes

to bring his work up to date. He writes of how he first came to own each marble, but he does not know where any of his marbles came from originally. When he asks his parents where marbles come from, they say that someone probably makes them in a big factory in England. There are sometimes a few marbles for sale in Coles or Woolworths in Bassett but these are brittle inferior kinds, and dozens of them have identical colors. True durable marbles are never seen in shops in Bassett. The few thousands of them that circulate among the boys of Bassett, wherever they may have come from originally, may never be replaced or added to, so that whenever a marble is lost it diminishes the total of marbles that a boy might collect in his lifetime, although years later another boy may find it again after a night of heavy rain has exposed it gleaming among the gravel and clay. One day Clement sees in a National Geographic magazine, in an article entitled *West Virginia: Treasure Chest of Industry*, a picture of a girl in a factory somewhere among steep forested hills almost as secretive as Idaho's [*another recurring object of Clement's fantasies —JB*] packing thousands of coloured marbles into small bags. He looks closely at the marbles. Many of them seem to be the cheap almost wholly transparent kind that are sold in shops, but a few of them are very like some of the precious richly-coloured marbles that cannot be bought anywhere in Australia, so that he goes on hoping for a long time afterwards that someone in those dark treasure-laden hills will send across the sea some of the true marbles that are still being made after all and save them from dying out altogether in Australia. *No one is even certain of the true names of the different kinds of marbles.* Clement owns a few of each of the kinds that he calls peb, realie, dub, catseye, bullseye, bott, taw, rainbow, pearlie, glass-eye, and chinaman, but other boys often use these names to describe quite different kinds of marbles. And on the few occasions when his father takes an interest in the boy's marbles, Augustine uses familiar names wrongly or calls some marbles by names that Clement has never heard before. Augustine himself first made Clement interested in marbles when he took out of an old tobacco tin that he kept in his wardrobe about a dozen marbles that he said were older than himself because he had found them somewhere around the old house at Kurringbar when he was a little boy. *Clement values these marbles so much that he never takes them out of the house and seldom shows them to another boy.* The next most precious are the dozen or so that he has found in the backyard and that must have belonged to the boy Silverstone who lived at 42 Leslie Street before the Killeatons moved there. Some of the rest he gets by swapping cards from packets of breakfast foods or toys and scraps of timber or coloured paper that his father brings home from the mental asylum where the patients make gifts for children in orphanages, or by sitting up straight in school or knowing his work on days when his teacher tidies her desk and finds there marbles that she has taken weeks before from some boy who was fiddling with them, or from older boys who think marbles are only for babies, and one day Mrs Riordan gives him a bag of marbles because he is always so well behaved when he visits her house. He watches the boys at school playing ring games for keeps, but dares not join in. The only marbles that he takes to school are a few inferior kinds which he tries to swap for some that attract him in other boys' collections. *He wonders how some boys can lose half a dozen choice marbles during a single playtime and not seem worried about it.*

Decluttering and aspirational minimalism are, of course, phenomena of affluence, though often of an affluence that is mixed with precariousness in ways that defy conventional measures of economic well-being. In some cases, adherents of aspirational minimalism seem to be participants in a cultural affluence that has become delaminated from market power. But the skillful management of stuff, the expenditure of considerable, seemingly unproductive time and attention on the acquisition, curation, retention, and redistribution of material things, is a more widespread phenomenon, one that cuts across differences in strategies of production, ideologies of property and personhood, and the scale of society in the sense of the discussion at the end of chapter 4. Indeed, games with stuff—institutionalized, self-recreating social phenomena in which inert if often sensuously appealing tokens become imbued with relational statuses—represent a candidate property of a distinctively human kind of sociality. Cognitive archaeologists sometimes point to beads—a covering term for all manner of ornamental tokens whose role appears to be symbolic as opposed to productive in the sense of contributing to meeting the community's food energy needs—as the earliest unequivocal, archaeologically legible signs of the kind of symbol use and interiority we associate with humans. But beads in isolation tell us little about the qualities of the symbolic roles, the meanings, these beads had for their makers and holders. More telling are signs of the *exchange* of beads—say, when daubs of ochre, bundles of plumage, or bits of stone inscribed with geometric designs turn up hundreds of kilometers from where the raw materials were available. Stuff glues us together.

What, then, is a crisis of stuff? Consider a historical parable: the Atlantic slave trade. Between the sixteenth and nineteenth centuries, the European maritime powers—the Portuguese and Spanish, succeeded, in turn, by the Dutch, French, and English—imported enormous quantities of goods to West Africa, from Senegambia down along the Bights of Benin and Biafra on what became known as the Gold Coast and Slave Coast. The goods in question were not selected with an eye toward maximizing comparative advantage, that is, exporting those things for which the European countries had more advanced manufacturing industries than their West African counterparts. Indeed, in many cases—textiles, glass, worked metal—West African states were more than capable of manufacturing high-quality finished goods both for everyday domestic use and luxury consumption, with a surplus available for export. Rather, the goods European traders were most avid to shift into West Africa were those, notably cowrie shells, bolts of cloth, and the copper armbands known as manilas, that were already most widely used in West Africa as currencies. The transformation of a regional slave trade—often a trade in captives and pawns rather than hereditary slaves, but a slave trade nonetheless—into a transoceanic slave trade was founded on a strategy of deliberate, partly coercive currency devaluation, with hyperinflation and concomitant political destabilization, not to say West African elites' growing taste for foreign prestige goods, contributing to the forcible exodus of Africans to Brazil, the Caribbean, and North America, an exodus whose legacy continues to shape our world.

It would be hopelessly glib to compare the flood of cheap consumer goods into the lives of rich and poor alike over the past couple generations to the deliberate currency devaluations of the Atlantic slave trade. For one thing, the operative regime of coercion is different—we need to be cautious about reading purposive malfeasance into the actions, say, of fast-fashion marketers the way we might into the actions of certain European traders and their state and joint-stock sponsors in West Africa three or four hundred years ago. But perhaps one dimension of what I have called the crisis of stuff is simply this: as humans, we are compelled to play games with stuff—it is a big part of how we fashion ourselves as social persons, as relational beings. Today, there is too much stuff in circulation, and so our games have taken on a centrifugal, destabilizing character.

Part of our new, or newly salient, concern to rationalize and contain the unruly purchase stuff has on our lives stems from the fact that stuff itself is increasingly unruly, and a growing part of the world of stuff participates in our bodily lives in an intimate, constitutive way. We are, in turn, given to taking an increasingly meronomic—concerned with part-whole relationships—view of our bodies.

I see now that part of what I have wanted to do in this book is to gesture toward a behavioral ecology of precariousness. But to do that, we must not just say something about the economic circumstances, in the conventional sense, that inform our relationship to stuff. We must equally say something about the broader network of biospheric circumstances of which these economic circumstances form one part.

When I wrote, just now, that *a growing part of the world of stuff participates in our bodily lives in an intimate, constitutive way,* I did not have in mind something abstract or metaphorical. I had in mind the way that one broad category of historically novel stuff, plastics, has been incorporated into our bodies. Many, perhaps most readers will be familiar with the Great Pacific Garbage Patch (GPGP), a gyroform concentration of synthetic polymer debris centered

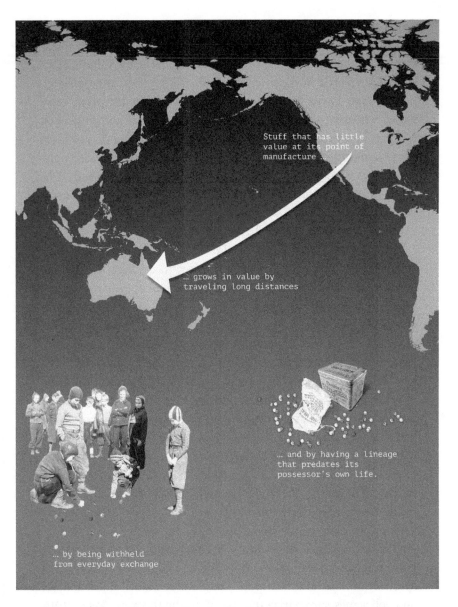

Stuff that has little value at its point of manufacture ...

... grows in value by traveling long distances

... and by having a lineage that predates its possessor's own life.

... by being withheld from everyday exchange

Figure 7. Games with stuff, as imagined by Gerald Murnane, *Tamarisk Row*. Illustration by Yoshi Sodeoka.

roughly halfway between California and Hawai'i. A series of aerial and surface surveys conducted in 2015 and 2016 afford some precision about the composition of the GPGP. At this time some 46 percent of the plastic in the gyre came from fishing nets. Microplastics, polymer specimens with a diameter of 0.05–0.5 centimeters, comprised 8 percent of the total plastic sampled by mass and 94 percent by specimen count. Within the area surveyed, synthetic polymers represented the overwhelming genre of marine debris, with polyethylene and polypropylene the most common types of plastic. The 2015–16 study estimated that the 1.6 million-square-kilometer survey region held some 78,400 metric tons of synthetic polymer debris across all size classes and polymer types.

The GPGP is not well bounded in time and space. It varies seasonally and interannually and is subject to a range of extrinsic forcings, including wind, seasonal ocean circulatory patterns, and infraannual meteorological phenomena such as the El Niño Southern Oscillation. The word *patch* is a bit misleading, because it is not as if, flying out over the ocean, you suddenly encounter a large contiguous region where plastic debris occupies the bulk of the ocean surface. Indeed, not all the debris has a specific gravity less than that of water, and it remains unclear to what extent the patch is really a circulatory column extending well beneath the ocean's surface. The role that William Gibson envisioned for the GPGP in *The Peripheral* (2014)—a kind of floating freeport used as a staging ground, variously, for installation art and terrorism—belongs, for the foreseeable future, to fantasy. Nonetheless, the GPGP has become an emblem of the remarkable degree to which plastics have, over the past two generations, become essential to how we live.

It is only in the past seventy years that synthetic polymers have become a salient feature of our world. In a landmark review of the "production, use, and fate of all plastics ever made," Roland Geyer and colleagues summarize the rise of plastic this way: "As of 2015, approximately 6300 Mt [million metric tons] of plastic waste had been generated, around 9% of which had been recycled, 12% was incinerated, and 79% was accumulated in land fills or the natural environment." More than half the new plastic ever produced has entered the world since 2005. Much of it exits its "use phase" within a year of manufacture—more than 40 percent of new plastic is single-use packaging for consumer packaged goods (aka "fast-moving" consumer goods), and most of this packaging has an expected use-life of less than a year. Until recently, the bulk of that part of decommissioned plastic—again, mainly consumer packaging—not incinerated, sent to landfills, or simply dumped has gone to China. Between 1992 and 2017 China imported some 45 percent of all plastic waste segregated for recycling in other parts of the world. In December 2017, China announced new restrictions on the import of hazardous waste that amounted to a ban on the import of postconsumer plastic, calling into question the viability of the plastic recycling schemes that had grown up in Europe, North America, Japan, and Oceania over the preceding thirty years.

Though the vast majority of commercial plastic is not designed to biodegrade, it does, of course, disintegrate over time, particularly in sunlight. This is how it comes to be, for instance, that a significant proportion of the plastic in the Great Pacific Garbage Patch occurs in specimens of less than half a centimeter in diameter. Plastic at this scale and smaller warrants consideration as a new category of persistent organic pollutant, though its toxicology remains yet to be elaborated. In some instances, plastic debris seems to serve as a vector for organic and heavy-metal pollutants, amplifying the rate of release of these pollutants in the gastrointestinal tracts of endotherms by a factor of up to thirty. Evidence of bioaccumulation—the reconcentration of plastic as it makes its way up the marine trophic ladder from gastropods and other sediment-bed filter feeders to fish to birds, pinnipeds and cetaceans, and humans—is mixed, and patterns of trophic transfer seem to vary with the type of synthetic polymer in question. What is not in question is that the body burden of synthetic polymers and associated toxins, among forms of life farther along the trophic ladder from primary production (that is, autotrophy, as in phytoplankton), is increasing.

What has received less attention than the rapid biospheric accumulation of plastic is the fact that the trophic network through which that accumulation unfolds is itself beset with a certain precariousness. This is something that it is difficult to get a synoptic view of, in part because we have not had a good baseline for the taxonomic composition of the Earth's biomass. One recent effort to formulate such a baseline estimates, not surprisingly, that plants make up the vast majority, by mass, of the Earth's living things, at 450 billion metric tons of carbon, followed by bacteria (70 gigatons of carbon), fungi (12 gigatons of carbon), archaea (7 gigatons of carbon), animals (2 gigatons of carbon), and viruses (0.2 gigatons of carbon). Within the animals, one half of the two gigatons of carbon consists of arthropods, with finfish responsible for another 0.7 Gt C. Wild birds and wild mammals, in this census, represent a negligible contribution by carbon mass, at 0.002 and 0.007 gigatons respectively. Humans, as a population, are an order of magnitude more massive (0.06 gigatons of carbon), with the carbon mass of domesticated livestock nearly double that of humans. That is, as the authors note, "humans and livestock outweigh all [other] vertebrates combined, with the exception of fish."

We have become accustomed to attending to the trophic power of that which is ordinarily invisible—the microbiota that inhabit the surfaces and orifices of our bodies and the length of our gastrointestinal tract, to take one example that has been particularly salient lately. But selective pressures cascade down the trophic network just as they filter up, and the loss of consumers at the top of the trophic network represents a threat to the integrity of the whole comparable to that of the loss of producers nearer the "primary" (photosynthetic, chemosynthetic) end of the spectrum. One team of investigators has been moved to describe a "trophic downgrading" of the biosphere, with the loss of apex consumers, carnivorous as well as herbivorous, prompting rapid reorganizations of biome structure. In some instances, large herbivores consume dry biomass that would otherwise serve as fuel for wildfires. Fire regime, in turn, as we saw in earlier chapters, has a range of cascading implications for vegetative structure and the economic strategies available to large animals, humans among them. In other cases, predators consume large animals, including primates, that otherwise serve as reservoirs of zoonotic parasites, with implications for the health of neighboring human populations. In freshwater and marine settings, the presence or absence of predatory finfish cascades down to inflect the relative abundance of zooplankton and the phytoplankton they consume, with implications for the capacity of the body of water they inhabit to serve as a CO_2 sink.

One of the difficulties of formulating long-term predictions about high-dimensional phenomena such as trophic cascades is that you cannot run history over and over to see what happens (though you can simulate it). But in the case of how the removal of megafauna from an environment affects its biome structure, we do have historical material to work with. The late Pleistocene and early Holocene saw rapid megafauna extinctions in the Americas, Australia, and Europe. There is a long-running debate on the degree to which humans were responsible for these extinctions. Most of the scenarios for a human role come down to some kind of *overkill*, that is, the hunting of large animals to the point where their populations were no longer self-sustaining. My own view is that overkill scenarios are somewhat wanting, both empirically (a want of evidence in the form of faunal assemblages) and conceptually (from what we know of human foraging behavior, it seems unlikely that specialization in game too large to be hauled back to camp would be widespread)— but that humans were doubtless implicated in megafauna extinctions in ways that are archaeologically and conceptually less legible than simply overhunting (recall our caveats, in chapters 2 and 3, about the episodic bias in our reconstruction of the past). Regardless of the causes of megafauna extinctions, we can say something about the consequences. Large herbivores, notably the extinct proboscideans we associate with the Last Glacial Maximum—mammoths, mastodons—along with ground sloths, beavers, and assorted large ruminants, exerted broad influence on the relative abundance of mature arborescent vegetation, with implications in turn for fire regime—fire here represents an "abiotic herbivore" with a role complementary to that of herbivorous fauna. The disappearance of these large consumers and destroyers of woody vegetation precipitated, depending on climate and soil, a transition to biome states governed either by fire or by "bottom-up constraints" such as water potential and soil nutrient status.

The cause of the contemporary disappearance of apex consumers is multifactorial but significantly, if not predominantly, human in origin. The appropriation of land for livestock and human settlement plays a key role. So does the rising incidence of periods of elevated temperature, which are stressful in their own right but also contribute to the incidence of zoonotic infections such as COVID-19 (it remains an open question how and to what degree climate change figured in the pandemic that began in 2020—dense coresidentiality of humans and other gregarious vertebrates was by far the more significant factor). Trophic truncation plays out slowly, via the reduced fertility and coping alluded to in chapter 4. But it also plays out in highly visible episodes of mass mortality. These have been observed lately across a wide range of animal taxa—insects, birds, starfish, fish, amphibians, mammals. To judge from references to mass mortality events in the journal literature, both the incidence and magnitude of these events are increasing over time.

Trophic truncation is one dimension of a broader reorganization of the dynamic landscape that characterizes chemical activity, including biological activity, at the surface of the Earth. In the language we used in chapter 4, we could characterize this reorganization as a shift away from the basin of attraction that has served as the theater of our own trophic ascent. Many of the salient near-term features of this reorganization appear, from a human perspective, to be shifts away from stability and toward volatility—think of the rising incidence of coastal flooding forecast for the coming decades, or how the decline of polar sea-ice cover serves both to amplify the strength of midlatitude cyclonic storms and make the incidence of such storms more variable. But in fact the reorganized Earth system will likely be more stable than the regime we have known in the sense that it will embody a deeper basin of attraction—it will be more resistant to perturbance, whether in the form of extrinsic forcings (orbital, volcanic)—or in the form of carbon sequestration or other climate-mitigation interventions undertaken by humans.

TREADMILLS AND SKILLS RESERVOIRS, AGAIN

Early on, recall, we were concerned with the cumulative quality of human culture. In the views of the trait transmission theorists whose work we canvassed in chapters 1 and 2, it was cumulativity that distinguished the role of socially learned skill in human adaptive strategy from its role in the adaptive strategy of other vertebrates. Corvids, cetaceans, and other-than-human primates acquire behaviors by observing others and formulate novel behaviors by independent tinkering. But as a rule, social learning and independent innovation do not build on one another to form a fabric of interwoven recurring partial behaviors that persists and grows over time—save in humans. It is the fabric-like quality of culture in humans, the fact that behavior acquired by different strategies and in different domains—about what is good to eat, say, or how to keep fire on hand for use at some future point in time—become mutually implicated, that has made it so integral to human behavior. By the same token, the threat that the fabric might unravel for want of enough good models for social learning presents a risk to the human strategy not present in other culture-using animals. Thus the image of the demographic treadmill: we are forever running to keep up with the cognitive and social demands of reweaving the cultural fabric.

I expressed skepticism about how trait-transmission theorists formulated their treadmill. This skepticism took two forms. One was that their somewhat mechanical view of the social transmission of skill does not correspond either to what has been observed of childhood development in subsistence societies or to what we know of how kinesthetic empathy functions in the enaction of coordinate behavior. The other was that the traits they emphasize represent a limited subset of recurring behavior, that which is either archaeologically salient or tractable to episodic imagination, and that this subset is not representative of the range of strategies adopted by human communities in different times and places. I was less skeptical that there is something distinctive, in comparison with other animals, about the degree to which human culture is cumulative. This observation feels worth holding on to. But it warrants qualification.

For one thing, cumulativity is not the only thing that distinguishes the human engagement with the world—the world of others and the world of stuff—from that of other animals. There is, for instance, the strange way humans imitate others, which suggests that, from an early age, we seek to reproduce what we infer to be the distal intent of our observational models as opposed to the motor procedures they adopt to achieve that intent (say, retrieving something from a high shelf, as opposed to knocking a broom handle against it—we are quick at grasping the purpose of the motor manipulation rather than simply replicating the manipulation itself). There is reflexivity and self-awareness and, related to both, the capacity to imaginatively recompose memories to generate counterfactual past and future scenarios. Other animals recognize themselves in mirrors, but they do not reflect on what might have been had they made different choices in the past. There is the syntactic sequencing of recurring partial behaviors into operatory chains that demand an awareness of settings and events that transcend the episodic horizon—a quarry that can only be visited at certain times of the year, say. There is the formulation of symbolic registers, as with the distinctions we draw among different kinds of kin, different kinds of persons, and different kinds of living things. There is, simply, the satisfaction we get from acting in a coordinate way with others.

All these things are found, in varying degrees, in many living things, and the way they have come together in humans may not prove particularly adaptive in the long run. My point in enumerating them is that these are all more fundamental properties of social cognition, as it unfolds in humans, than is cumulativity. The sense in which they are more fundamental is that cumulativity entails these things, but the converse is not true. I try not to concern myself too much with ontological parsimony—with excluding from our lexicon any concept that is not strictly necessary to explain some phenomenon of interest, in this case the persistence of culture. But perhaps we would do well to conjure up images that speak to the underlying causes of cumulativity. Perhaps culture abides not on a treadmill but on a dance floor—that is, through the continued enaction of a coordinate pattern of activity that, through whatever chance of evolution has become its own reward.

For another thing, cumulativity, as must be clear by now, is not a straightforward accumulation but rather an ongoing process of substitution and deferral—and forgetting. These take a number of forms. As an example of one, consider the shift from analog to digital telephony, described by musician and essayist Damon Krukowski in his book *The New Analog*. The carbon microphones used in analog telephone handsets were characterized by a broad but uneven frequency response. Unevenness of frequency response is a characteristic, to one degree or another, of practically all audio signal processing equipment. It is, for instance, why the world sounds not just quieter but muffled or distorted when you wear earplugs (though in fact, you can get earplugs designed to offer a flat frequency response, and I recommend them). But what analog telephony lacked in fidelity it made up for in indexicality. That is, certain features of the way the world sounded through an analog phone signal indexed or referred back to properties of the environment where the sound was coming from in ways that lent the sound a quality of copresence. You could hear, for instance, when a speaker was holding the mouthpiece directly against her mouth, a phenomenon known as proximity effect. You could hear background noise that gave you a sense of where your interlocutor was.

Today, these things are gone from a phone conversation even though the microphones are of higher quality and the transmission is no longer subject to the vagaries of copper wire. This is a product of a series of design decisions on the part of mobile telephony handset manufacturers. Among these, handsets today use multiple microphones to achieve source separation and noise cancellation, isolating the speaker's voice from neighboring sound and transmitting just the voice signal. You can argue about the validity of the choices that informed the design of digital telephony. But however valid they may be, they have entailed a recalibration of our rubrics of audition and a forgetting of the rubrics that preceded them.

Perhaps this is a trivial example—though one, I would argue, that is symptomatic of a broader disengagement from the acoustic world that, as we saw in chapter 4, may prove maladaptive. Substitutions of this sort are value-laden, and I do not wish to argue that the rubrics of sensory attunement that, by virtue of acculturation, feel comfortable to me are better than those that have lately become widespread. But substitution is not the only way in which technological evolution is something other than cumulative. A more powerful and unnerving form of cultural change is *deferral*. Here I have in mind a topological deferral, a shifting of the skills implicated in some high-level task—feeding oneself, say—away from one's body and out into a network of activity whose behavior is only loosely coupled to one's own. In a way, deferral is a natural evolution of coordination. Media of deferred coordination, such as money, allow a society to extend its fabric of socially conveyed behavior over a broader horizon of time and space. It is no longer the case that every individual must acquire all the skills implicated, say, in procuring food. Those aspects of food procurement that are more distant in time and space from the act of eating gradually fade into the background save for a small number of specialists.

This may seem like an obvious point. I make it simply to show, again, that the argument from the first half of the book is relevant to our own lives, notwithstanding the substantial elaboration in the technological scaffold of human behavior in the time since the events described there.

In chapter 2 I proposed that how we gauge the adaptive value of different strategies for getting on in life depends on the time horizon over which we imagine those strategies unfolding. Behavior that might seem counterproductive or retrograde in the near term could turn out, in the fullness of time, to represent the maintenance of a skills reservoir useful under extreme circumstances that do not occur very often. Another way to look at the horizon of adaptive value would be in terms of the frequency of extreme events. As this frequency increases, so does the adaptive value of go-for-broke or windfall strategies.

Perhaps this is another part of what I had in mind by *a behavioral ecology of precariousness*. One reason one might be inclined to turn away from the technological scaffolding of one's own life is a sense that that form of scaffolding is brittle in the face of volatility, or at least, less supple than the enactive form.

In *The Road*, Cormac McCarthy reminds us that before the road is a place of freedom it is a place of lawlessness and danger. Or perhaps this is a matter not of before and after but of entitlement. Some people, even when they lack, or have lost, the entitlements that come with property, cash flow, credit, and professional socialization, retain those that attach to privileged statuses of ethnicity, gender, nationality, or urbanicity. (Some people have neither kind of entitlement: you would not call the lives of people in these circumstances

precarious—you would simply call them crushing.) These entitlements have an inertia. Their loss and gain is subject to hysteresis or
lag—they lag the entitlements of stuff, sometimes by generations.
These other-than-stuff entitlements provide a buffer against the accelerationist churn, to return to the evocative if opaque phrase I used in
my first notes for this book a year ago, that renders socialization—
upbringing, education, work life—less a developmental investment in
adult responsibilities along the lines we looked at in chapters 1, 2, and
3, more the first in a series of manic episodes of self-invention.

Sometimes the lawlessness of the road is deliberate, and sometimes it is the product of studied indifference but then becomes
useful to certain parties, as in the extralegal detention of frontier-
crossers by states that represent themselves as paragons of the rule of
law. In *The Road*, McCarthy imagines life in what had been the eastern United States, ten years after catastrophe has reduced civilization, as it were, to ashes. (The catastrophe remains unspecified, but it
could be nuclear war—the protagonist sees a flash of light.) A father
and son make their way down through Appalachia and across the
Carolina Piedmont to the coast. Their hope is not so different from
that which might lead you or me to go to the sea. They, and we, hope
that the climate will be more salubrious, the light clearer, the people
kinder, at the edge of the sea. For the protagonists of *The Road*, these
hopes prove mostly unfounded. The father tells the son that they are
"carrying the fire," by which he means that they are the good guys, the
ones who, whatever they might have to do to survive, have not lost
their moral bearings. There are certain things they will not do, even
though these things have become common: they will not enslave
others, they will not kill and eat babies. This may sound like a low
bar, but there you go. Though they are carrying the fire, many nights
they have no fire, either for warmth or cooking—because they have
run out of fuel, or have had to abandon their stove, or the army of one
or another petty warlord is on the road and they cannot risk being
discovered in their bivvy in a hollow in the forest. Curating fire has

become their metaphor for order and purpose, and for a mode of society that does not rest on exploitation.

Over the course of the period in geological history known as the Ordovician—488 to 444 million years ago (Ma)—the Earth experienced a cooling pulse marked, after 460 Ma, by episodes of glaciation. This despite the fact that atmospheric concentrations of CO_2 were an order of magnitude greater than at present and the mean temperature was over 21 degrees Celsius. The onset of the glacial cycle appears to have been preceded by a precipitous drop in CO_2 concentration, from some fourteen to twenty-two times the present atmospheric level (PAL) to a more modest eight PAL and a mean temperature of 17 degrees Celsius. It is difficult to draw comparisons between the Earth system of the Ordovician and that of our own day, because the distribution of terrestrial mass at the surface of the oceans was dramatically different than the distribution of continental land masses as we know it, with implications for ocean circulatory phenomena; the exchange of gases between the lithosphere, atmosphere, and ocean; the surface albedo (solar reflectance) of the Earth; and so on. Still, it is interesting to ask what could have caused the CO_2 drawdown we observe in the climatological record of the late Ordovician. One hypothesis is that a key role was played by the earliest terrestrial plants.

These were avascular plants, similar to the bryophytes—mosses, hornworts, and liverworts—that feature in the understory of damp biomes today, along with lichens, a class of holobionts composed of algal or cyanobacterial colonies scaffolded on fungi. Lacking the endogenous vascular structure of tracheophytes, bryophytes and lichens depend on some mineral or ligneous substrate to provide both a physical scaffold and a growth medium. But in the Ordovician, there were no ligneous vascular plants, nor was there much of a humus. Terrestrial surfaces consisted of bare rock, notably basalts, newly erupted from the Earth's magma. The minerals in this rock—phosphorus, magnesium, iron, calcium, potassium—provided the nutrition for the growth of bryophytes and lichens. Plant life depends on plants' evolved strategies for accelerating the weathering of silicate mineral substrates, in the process drawing carbon out of the atmosphere and sequestering it in the ground. Experimental and simulation studies suggest enhanced weathering by avascular plants might have been sufficient to tip the Earth's climate into a glacial cycle. Recently, enhanced weathering has been suggested as a strategy by which humans might intervene in the carbon cycle to slow climate change.

It is important to maintain a sense of scale. The "accelerated" drawdowns of atmospheric carbon that characterized the second half of the Ordovician unfolded over periods on the order of a million years. By way of comparison, the hominin clade appeared less than three million years ago, anatomically recent humans—Neanderthal, *sapiens*—after 500 ka. We might go so far as to say that social acceleration, as we experience it today, is but one aspect of a process of the acceleration of the carbon cycle. Perhaps acceleration is, in some deep if difficult to characterize way, in the nature of the evolutionary path we are on. In the long run we represent a transient reversal in the cycling of matter through states of increasing entropy—entropy is the treadmill you cannot outrun. In the short run, we would do well to recognize that the design space of cultural evolution is not given but fashioned. Its fashioning stems more from the coordinate enaction of sensorimotoric presence than from the material precipitates—the stuff—of that enaction. It may be too late—politically, technologically, chemically—to halt the flooding of the strait between the human past and the human future. Either way, our adaptation to that future begins not with technology but with physiology.

Postscript

Initially, recall, I had no intention of writing a preface. I wanted to open with a couple lines of high-minded text encouraging the reader, as it were, to enter this book as one would enter a river—here I imagine something broad and slow-moving, Amazonian, like a sea save for a subtle anisotropy in its currents, the water turbid, lit by the moon, salt marsh extending to the horizon in either direction—to allow oneself to be borne along, the topography and destination coming into view gradually. Perhaps this was a temporizing strategy, a way of buying myself time, for I did not know, when I started out, eleven months ago today, where I would end up. I had themes I wanted to touch on—the strategic trading-off between material and enactive practices, the difference between survival and flourishing—and episodes I wanted to visit—peri-Holocene Tasmania, pericontact Leeuwin-Naturaliste, Onyuudani at the onset of a late typhoon season. Other elements became salient as I went along. I wrote the proposal that landed me the opportunity to write this book under the heading *The Scaffold*, and de Finetti diagrams of the adaptive landscape of rock-paper-scissors games (figure 8) occupied my field of vision all through the month just before I started writing. But the tension between landscape and scaffold as devices for reasoning about evolution did not come into focus until I had finished the first draft.

Just as I did not intend to write a preface, so I did not intend to write a postscript. The preface came about in response to concerns from early readers that I had not made clear the stakes for me in writing this book—who I was, how I'd come to find myself wedged between divergent worldviews. The postscript has come about in response to a related concern: that I have not made clear the broader stakes, the stakes for a theory of what it is to be human, the stakes for anthropology.

Writing this last phrase gives me a cold shudder. If there is one river I have no interest in wading into, it is that of disciplinary boundaries, a torpid place of schistosomes and flesh-eating fish. But here we are. Software designers speak of a *minimal viable product*, something with the minimum set of features needed to attract and retain users without introducing excess risk into the design and implementation. In the spirit of a minimal viable product, I offered a minimal viable definition of anthropology: *the study of how culture mediates human adaptation to environment, with emphasis on the ecological determinants of behavior, the coevolution of individual, community, and milieu, and the nonlinear interaction of phenomena unfolding over time scales of ten milliseconds to one million years.* This definition says nothing that would privilege analytic or interpretive approaches, let alone a particular view of the appropriate relationships between observation, model building, and emancipatory politics. But clearly, I have opinions on these matters, and these opinions inform what I've written. And I hinted at something more grandiose: But really, *more than one colleague has said to me,* you're trying to create a new discipline. *Indeed, it might be something you could call* sensorimotor ecology—*or, extending the project beyond sensory and motoric behavior in the conventional senses,* semiokinetic ecology.

In the spirit of boro, I won't try to sew everything up. At various points in this book, I've expressed dissatisfaction with the theory of knowledge operative in the texts under discussion. I've tried to keep this stuff to a minimum. Still, to some readers it must have been mystifying, to others wearying. Let's consider some highlights.

From chapter 1: *The assumptions built into the treadmill theory are value-laden, which is to say that they refer, ultimately, not to things that can be arbitrated on a factual basis, such as the relative significance of population size as opposed to environmental risk in structuring foraging strategy or even the best way to measure technological complexity, but to questions such as How should we live? What is a good life? What is flourishing?*

From chapter 2: *Skill transmission models, lacking an explicit theory of how the environment, material and social, scaffolds behavior over time, cannot contend with variation in degree of selection and niche breadth. More broadly, they cannot contend with how the relative merits of skillfulness and technological complexity are conditioned by somatic coordination, both in the course of the exercise of skill (say, on a hunt) and in the subsequent redistribution of the goods gained by the exercise of skill (food, building materials, etc.).*

Chapter 3: *Over the past three chapters I have tried to trouble our assumptions about the relationship between technology and economic strategy. Indeed, I have sought to broaden our view of strategy to include patterned behavior that is not willful in the conventional sense*—conducted with an awareness of scale—*but rather the contingent product of a chain of gestures none of which anticipates the distal outcomes—just as, when we speak, we do not anticipate how our ephemeral gestures will contribute to the emergence of a novel speech register. Niche construction is not like running on a treadmill. It is, rather, like negotiating a rugged terrain.*

It's in chapter 4 that I really let myself go. From the discussion of behavior sampling strategies: *My own view, based on my experience designing prototype systems for the remote sensing of mood, alertness, and social stress, is that the richest way to incorporate remote sensing into social science is by using it to enhance the degree to which participants' capacity to reflect on their own states of being gets folded into the data. This reflexive capacity is not the problem it is often made out to be. The greater problem is the methodological blindness that comes from imagining we could devise value-neutral measures of behavior.* Not to say most of the section on "social complexity."

The low-hanging fruit would be to characterize these asides as a running plea for some kind of middle ground between the analytic and the interpretive, the reductive and the constructive. I said at the outset that I do not think modeling—dimensionality reduction, the selection of certain features of a phenomenon to focus on to the exclusion of others—is incompatible with epistemological humility. Indeed, chapters 1 and 2 represent an extended argument that for models to be robust they must be humble—they must evince a certain boro quality, a patchiness, together with a cautiousness about the adequacy both of our evidence of human behavior in the wild and of our strategies for making sense of that evidence.

But there is something more specific that links the passages cited above. This something consists in at least three overlapping parts: multitemporality, semiokinetic ecology, and the individual as emergent phenomenon. These warrant brief exposition.

MULTIPLE TIME HORIZONS

One of the recurring themes of this book is that niche construction unfolds over multiple time horizons simultaneously—crudely, the episodic, the everyday, the biomic, and the somatic. A supple theory of human behavior, to say nothing of its environmental consequences, must take into account how selective pressures operating over multiple horizons interact. This is implicit, for instance, in the concept of a skills reservoir.

Skills reservoirs were nowhere in my thinking until the moment I wrote the phrase. It originated as a boro, a patching, of something left over from a previous book, *The Meat Question*, where, in a discussion of hunting as costly signaling, I left open the question of what, apart from a proverbially male obsession with ballistic display, could account for the differences in the risk profiles of men's and women's hunting strategies. It was when I began to write about how strategies of flourishing unfold over multiple horizons simultaneously that it occurred to me that in this instance competitive display might serve a function beyond signaling one's commitment to the community—that a windfall strategy, ordinarily low-yield but essential in periods of environmental stress, might disappear in extended periods of stability unless adolescents, or some subgroup of a community's adolescents, had an incentive to cultivate it independent of its day-to-day economic value.

No doubt this is a trivial observation, or at least an unoriginal one. But the concept of a skills reservoir feels useful, among other reasons, because it offers a corrective to the understanding of cumulativity that is folded into trait-transmission theories of cultural evolution. Part of what distinguishes human culture, you read over and over, from the culture of other primates, say, or dolphins, or crows, is that it is cumulative, it accrues over time, with new innovations building on previous ones. What makes the demographic treadmill a treadmill is that cumulativity is precarious, because the transmission of skill from one generation to the next is prone to errors. The paradigm case of error-prone transmission is the sound system of a language—kids hear things wrong, certain sounds are more likely to be misheard in certain ways, and so the phonology of a language changes over time in a characteristic, internally directed way—but one that is constrained by the biophysics of articulation and in any case is more or less neutral with respect to a community's chances of surviving and flourishing. By contrast, errors of transmission in economic strategy, the argument goes, are, as a rule, neither constrained

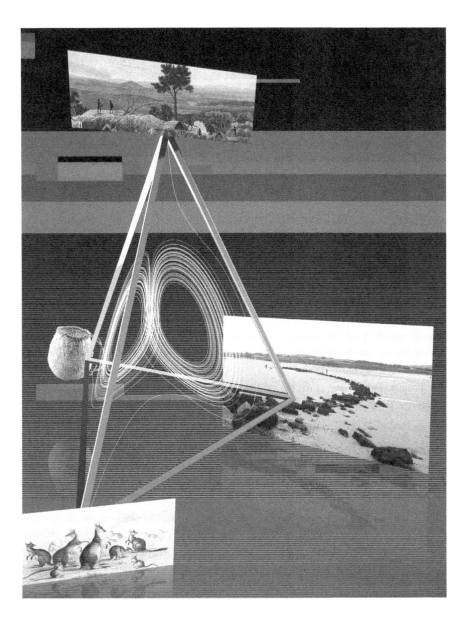

Figure 8. Skills reservoirs. Depicted is a phase diagram for a space of four complementary foraging strategies: *prescribed burning* (top); *gathering* plants, fungi, and insects (left); *trapping* fish in a weir (right); and *hunting* large, mobile vertebrates (bottom). Under evolutionary dynamics that are *continuous* (roughly: bounded first derivative, no sudden jumps in strategy distribution), where a strategy's growth rate is *positively correlated* with its benefit or return, where all *stationary points* in the dynamic are *Nash equilibria*, and where *optimal strategies not currently in use tend to be adopted*, "strictly dominated" (never optimal) strategies such as hunting will survive indefinitely—and thus be available when extrinsic risk makes them transiently optimal. A wide range of nonequilibrium evolutionary dynamics meet these conditions—but not the replicator and other pure-imitative dynamics favored by evolutionary game theory, which assume convergence to a stationary state.

Phase diagram adapted, with liberties, from figure 11 in Hofbauer and Sandholm 2011. Pyrogenic landscape detail from Robert Havell's 1834 print "Panoramic View of King George's Sound, Part of the Colony of Swan River" (present-day Perth), based on surveyor's sketches by Robert Dale. Medieval weir, Anglesey, Wales: Photo by Richerman, 2014, courtesy Wikimedia Commons. Kangaroo illustration adapted from Oliver Goldsmith, *A History of the Earth, and Animated Nature* (1774). Basket from a color engraving by Charles Alexandre Lesueur, Baudin Expedition, originally published in François Péron, *Voyage de découvertes aux terres australes, partie historique* (Paris: Imprimerie impériale, 1807–1816), plate 13. Illustration by Yoshi Sodeoka.

by the biophysics of specialized organs nor neutral with respect to a community's chances, so you end up with Tasmanians maladaptively shunning clothing and diving for abalone. For the reasons discussed in chapters 1 and 2, I find this argument not just unconvincing but conceptually crude, ethnographically ignorant, and numb to the bodily facts of experience. In chapter 5, I added that cumulativity is not really cumulative, or rather, it is cumulative in a shallow sense, cumulative in the sense of *being derived from in a path-dependent way* but not in the sense of *recapitulating the path in a way that would allow us to retrace our steps*. In fact, even *derived from in a path-dependent way* is too strong, since often, as in the trivial case of microphone technology, a novel technology represents a discontinuous divergence from a predecessor formulated to serve more or less the same function, and the construal of the novel technology as superior is loosely, and sometimes not at all, coupled to its success in filling that function.

Depending on how you look at it, skills reservoirs either represent an affront to cumulativity—the costly maintenance of a strategy that, be it evaluated in terms of energetic outlay, ease of acquisition, or what have you, is less effective than an alternative—or they represent true, recapitulative cumulativity, a cumulativity that makes it possible to return to "the old ways" when the environmental conditions that made successors attractive cease to obtain. But even to frame skills reservoirs in this way is to do them a disservice. In the case of gender-differentiated risk profiles in foraging, for instance, it is not that men's windfall strategies represent a fallback, to be dusted off in times of extrinsic stress—rather, they represent a strategy unfolding *concurrently* with the reliable, low-yield strategy more typical of women's foraging—but unfolding over a longer time horizon.

I must stress that in referring to windfall foraging strategies that present a high barrier to acquisition—years of ballistic conditioning, et cetera—as a skills reservoir, I am offering a hypothesis. To confirm or disconfirm this hypothesis, or, more precisely, to evaluate the degree to which the characterization fits the behavior, we would need long-term time series data about the effectiveness of different strategies under annual, decadal, and generational variation in environmental risk (for instance, prolonged drought or recurring inundation). Today, data of this type is difficult to get because foragers do have a fallback in the form of store-bought foods—so the role, if any, of windfall strategies as caloric buffers under extrinsic stress has been muted.

But the fact that skills reservoirs, not to say the precise intended sense of cumulativity, receive so little discussion in the literature on the selection and transmission of embodied skill canvassed in chapters 1 and 2 is a testament to the crudeness of the concepts of strategy and utility operative in this literature. Strategy and utility, as evolutionary theorists of behavior use them today, represent an inheritance from decision theory as it was formulated in the Cold War environment of the last century. Decision theory had no need of concepts of strategy and utility supple enough to model how behavior unfolds over transgenerational horizons, let alone over multiple time horizons concurrently. We do.

SEMIOKINETIC ECOLOGY

Sensory ecology is the branch of ethology that aspires to make sense of animal sensemaking faculties. It is perhaps most developed among the bird people, and I have been guided by Graham Martin's stirring *The Sensory Ecology of Birds*. In incorporating the motoric dimension of behavior into the descriptor *sensorimotor ecology*, I acknowledge the insight, formulated by philosophers of cognition working in the enactive tradition, that sense-making arises through the contingent coordination of sensing and movement. A tree has a presence, a thereness, for me insofar as, apprehending it with the exteroceptive senses (vision, audition, olfaction, cutaneous tactition), I have a running awareness of how its palpable form would change were I to shift my body so that the tree and I stood in a different relationship in space. Similarly, an animate being has a distinctive quality of presence for me insofar as its autonomous movements relate to my own movements in a way that is either more or less determinate, more or less in line with my prior expectations about how enminded presences move. When the world and our presence in it have a causal coherence, a sense that things fit together in time and space, this is the outcome of a stitching together of many such coordinate instances of sensing and movement at the boundaries between self, other, and milieu. Indeed, it is through coordinate sensorimotor activity that we fashion these boundaries anew in every moment. (Recall my observation in chapter 3 about how fire troubles these boundaries.)

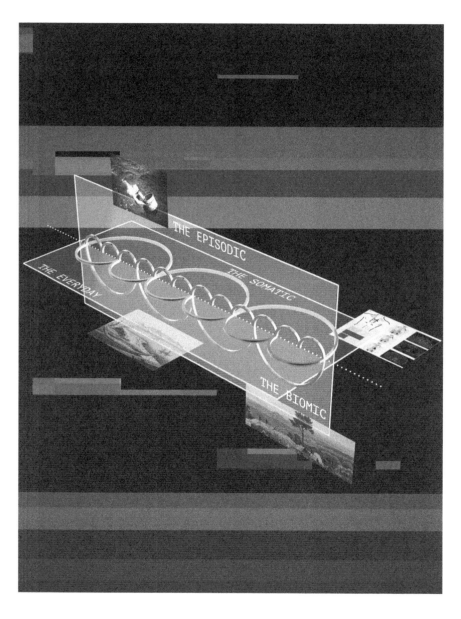

Figure 9. The multiple concurrent time horizons of niche construction. *Episodic* behavior such as breath-hold diving unfolds in dialogue with *everyday* and *biomic* behavior (shaping one's living environment to minimize the need for clothing) and *somatic* adaptation such as selection for thermogenic brown adipose tissue (BAT). (Note that BAT selection itself unfolds over multiple developmental and transgenerational horizons.)

Camp with bark-bundle canoes: engraving by Charles Alexandre Lesueur, Baudin Expedition, originally published in François Péron, *Voyage de découvertes aux terres australes, partie historique* (Paris: Imprimerie impériale, 1807–1816), plate 14. Pyrogenic landscape: detail from Robert Havell's 1834 print "Panoramic View of King George's Sound, Part of the Colony of Swan River" (present-day Perth), based on surveyor's sketches by Robert Dale. BAT PET imaging: Becker et al. 2016, figure 2. Ama-san (diver): Undated photo (c. 2005), courtesy Wikimedia Commons. Illustration by Yoshi Sodeoka.

By adopting the term *semiokinetic* I intend a modest expansion of the sensorimotor enterprise, one inspired, *inter alia*, by my colleague Hannah Landecker's observation that the same signaling cascades are implicated in olfaction—the sense of smell—and in the endocrine response to the presence of food in the gut. The former we think of as sensation, because it has a phenomenal dimension—when we smell something, there is something that it smells like. (This may be saying nothing more than that scents are associated with prior expectations, but this fact in itself opens up on the deep question of whether there is something to phenomenality that is irreducible to the "dispositional" or behavioral properties of physical stuff.) The latter is chemosemiotically similar to olfaction but ordinarily lacks that phenomenal dimension, and we have no name for it in everyday language. Analogously, voluntary muscular activity, or even involuntary muscular activity that interacts, in a determinate fashion, with voluntary muscular activity, as in the case of breathing, is motoric, whereas peristalsis, ventricular contraction, and other forms of central pattern-generated kinesis are not motoric in the ordinary sense of the word. But these other-than-motoric forms of kinesis are implicated in the coherence-making that unfolds at the contingent, shifting boundary between self and milieu just as much as those we think of as motor behavior. Indeed, from an evolutionary standpoint, central pattern generation, the internal coordination of metabolic and respiratory processes, likely preceded sensorimotor coordination as a driver of the emergence of central nervous systems—or the two systems, internal-coordinative and sensorimotor, may have developed independently for a time before fusing in a single nervous system. If, as I have suggested, what is distinctively anthropological is a concern with the ecological contexts of behavior, then anthropology needs an explicit theory of the semiokinetic interfaces between self, other, and milieu.

FOAMINESS, OR, THE INDIVIDUAL AS EPHEMERAL EMERGENT PHENOMENON

Indeed, it needs more: it needs an explicit theory, or at least a set of hypotheses, about how individuals emerge, in the evolutionary, developmental, and episodic senses, from a semiokinetic milieu that is prior to the individual just as the reaction-diffusion phenomenon mentioned near the end of chapter 4 is prior to the waves, loops, rings, and other persistent body-like forms that emerge, under certain parameter conditions, in reaction-diffusion systems. It would be easy to read this book as an argument against methodological individualism, that is, the tendency to take the human being, or, more precisely, the liberal subject, as the unit of analysis in the science of human behavior. This is a plausible interpretation but an incomplete one. For one thing, I don't feel that adherents of interpretive-constructive approaches to human behavior, for all that they rail against the tyranny of the liberal subject, have done any better than their analytic-reductive antagonists at formulating a workable alternative. At the end of the day, our experience as self-aware presences with bounded extension in space and, we hope, continuing extension in time exerts a strong pull on our intuitions about the constituent dimensions of the behavior of living things, human and otherwise. For another thing, my interest lies not in saying what does not work but in coming up with something new that might. We would all benefit, for instance, from a stronger awareness of the evolution of multicellularity, and in particular of the divergent strategies of multicellularity that took hold in the plant and animal domains. Indeed, an early outline for this book included a plan for a chapter with the working title "Foams" that would have offered just that.

What would a foamy theory of behavior look like? For one thing it would start with, and circle back to, the iterative, semiokinetic fashioning of boundaries between body and milieu. Initially this would feel alien, as the clumsiness of my account of the Noongar relationship with fire in chapter 3 suggests. With time, it is my hope that viewing semiokinetic coherence-making as prior to individual coherence-makers, not just in an evolutionary or developmental sense but as something stronger, methodologically—metaphysically?—prior, would open up a suppler option space for reasoning about behavior as it unfolds in the everyday, body-bounded way. If we learn to see embodiedness as a transient form of phase coupling in a semiokinetic milieu that precedes and exceeds us, perhaps we will be better equipped, as I hinted at the end of the preface, to see the design challenge of a world of exhaustible resources not as one of footprint reduction or sustainability or even circularity but of decoherence, of graceful disintegration.

But this may be a fantasy.

Glossary

Boro   "Tattered, patched." An ornamental style of textile design, descended from vernacular mending habits of Edo-era Japan. I have taken boro as a model for how to sew together the themes of this book.

Cumulativity   The phenomenon whereby culture builds over time, so that culture represents the products of long unbroken chains of intergenerational transmission. Often held to be the distinguishing feature of culture in humans as opposed, say, to crows, dolphins, or other-than-human primates. For critique, see chapter 5 and postscript. See also *treadmill*.

Enaction   A family of views in philosophy of mind that hold that cognition is not something that happens "in the brain" but rather is the product of an ongoing probing of the relationship between what we experience of the world, including our own bodies, through our senses and how we situate ourselves in relation to the world via movement. See also *semiokinetic milieu*.

Enregisterment   The process whereby recurring fragments of behavior become regimented, widespread in a community, and a signal of one's membership in that community. Consider, for instance, how a distinctive habit of assimilating certain vowel sounds and pronouncing r-final words functions in signaling fluency in Received Pronunciation (RP)

English: *In the wake of the fa [fire], the inhabitants of Holloway have been left without pa [power].* See also *metapragmatics.*

Epiphyte A plant that uses another plant as a scaffold without relying parasitically on its host for nutrition.

Extrinsic risk A factor tending to militate against survival and flourishing that is more or less independent of (extrinsic to) the behavior of the community in question. So, for instance, urbanization among humans represents an extrinsic risk for many kinds of birds. Orbital forcing of the sort that led to the cooling phase of the late Pleistocene represented an extrinsic risk to human communities in many parts of the world. For the human communities that stand to suffer the most under climate change over the next couple generations, climate change represents an extrinsic risk even though it is caused by humans—because it is not under their control (and, at this point, the same could be said of all of us, no matter how insulated we may be from climate change—the nonlinear lag in the response of climate to human behavior means we are no longer in a position to forestall the near- and medium-range effects of warming set in motion by the onset of industrialization, though more privileged communities are better positioned to adapt).

Foaminess, ethos of An approach to understanding our role in the world and the bases of human flourishing that starts with the view that individual selves, not to say embodied selfhood in general, represent ephemeral phenomena. See also *semiokinetic milieu.*

Game theory A family of mathematical formalisms that generalize the problem of optimization—solving simultaneously for the maxima or minima of multiple independent parameters—to situations where multiple actors, each with its own perspective and interests, are trying to maximize advantage simultaneously. In some situations, the structure of the game may encourage individuals to cooperate, in others to compete or even to undermine one another in spiteful ways, that is, in ways that hurt the ones doing the undermining as well as the ones being undermined. Evolutionary game theory is a subfamily of game theoretic formalisms that stipulates that individuals may revise their strategies over time to maximize long-term advantage in situations where they must repeatedly interact with other members of the population or, in some cases, members of other populations with a distinct range of strategic options. See also *landscape, strategy.*

Holocene The epoch in the Earth's geologic history dating to approximately 11,700 years ago and extending into the present day (as of mid-2020, the

International Commission on Stratigraphy had not recognized any of the proposed dates of onset for a proposed new epoch to be known as the Anthropocene). It is characterized by an overall warming trend relative to the Pleistocene, the epoch that preceded it, and, in certain human communities, by the efflorescence of agriculture and settled living. See also *Pleistocene*.

Image schema   A formalism in cognitive linguistics, philosophy of language, and neighboring disciplines for characterizing regularities in the relationship between tactile-kinesthetic experience and semantic patterning in language. Examples include oppositions such as those between slippery/smooth and rough and between inside and outside, along with concepts encoded in the temporal and spatial semantics of verbs, for instance passing over, under, or through, or starting, concluding, abiding, and occurring concurrently with. Some image schemata recur across widely divergent cultures in ways that suggest they have some basis in our innate faculties. See also *semantic template*.

Landscape   Used metaphorically, a conceptual template for making sense of the process of adaptation, for instance by genetic or cultural means, in a population. The dimensions of the landscape represent independent parameters in the space of possible sets of genotypes or phenotypes. An individual or population is said to occupy a position in the landscape given by the values it exhibits for those different parameters. See also *scaffold*.

Metapragmatics   The dimension of language use by which speakers continually comment on, enforce, and tactically diverge from enregistered norms of usage. One common example is the negotiation of forms of address (intimate, familiar, professional, deferential/honorific, etc.) between individuals whose relative social status is uncertain or contested. Another is codeswitching—as, say, when I conclude a definition of metapragmatics by saying, *All's I'm saying is, metapragmatics is important,* invoking the distinctive clitic complementizer *'s* and the cleft sentence structure of African American English to soften the imposing formality of what has come before and draw my listener/reader into an intimacy predicated on an unspoken shared understanding that the register of formal metalinguistic description can be off-putting.

Operatory chain   A formalism in cognitive archaeology for describing the relationships, concurrent and sequential, between steps in the performance of some recurring behavior such as the manufacture of a tool (in the case, say, of a stone blade or point, procuring raw materials, preparing roughs or blanks, undertaking fine work or retouch, firing, preparing resins and other hafting materials, preparing handles or shafts, hafting . . .).

Pleistocene   The geologic epoch preceding the Holocene, running roughly 2.6 million years ago to 11,700 years ago, characterized by increasing volatility in climate and terminating in a cool phase known as the Last Glacial Maximum. The Pleistocene coincides roughly with the rise of *Homo*, the genus that includes human beings.

Prescribed burning   The deliberate setting of fire in woodland or grassland for purposes ranging from driving out prey to stimulating the growth of pyrophytic (fire-adapted) vegetation.

Scaffold   A conceptual template common as a metaphor in philosophy of evolution, where it is used to evoke the role that a behavior pattern or extrinsic factor has in supporting and constraining the development of some behavior. So, for instance, it is common to hear statements to the effect that reliance on meat scaffolded the evolution of big-brainedness and thus much of higher cognitive function in humans—though, as I've argued in *The Meat Question*, this is wrong.

Scale   The question of when and how it is appropriate to draw inferences— about evolution, behavior, et cetera—across scales of time, space, and social complexity is among the most challenging methodological and epistemological questions for evolutionary theory. See discussion at the end of chapter 1, the end of chapter 4, and in the postscript.

Semantic template   An often implicit metaphoric ground for structuring relationships between pairs or across sets of lexicalized concepts, for instance uphill–downhill or with the tide–against the tide. See also *image schema.*

Semiokinetic milieu   The material ground for the ongoing processes of signaling and movement by which individual selves differentiate themselves from one another and their shared environment. It includes sensing and motoric behavior in the conventional senses along with a broader field of activity that is cognate, in its dispositional or physiological parameters, with sensory and motoric behavior but lacks the reflexive phenomenal dimension that we ordinarily associate with sensorimotor behavior. Examples of these broader activities would include chemosemiotic phenomena that unfold in the gut of vertebrates, including us, and involve some of the same signal transduction pathways as those implicated in olfaction (smelling)—but which we don't think of as sensing because they lack a phenomenal something-it-is-likeness.

Semiokinetic ecology is my proposed formulation for a way of investigating how culture mediates the individual and collective adaptation to milieu over time, not to say the role of the community in shaping that milieu (that is, niche construction). It emphasizes the ephemeral character of individual selves and the self–milieu distinction. See the postscript for discussion; see also *foaminess, ethos of.*

Skills reservoir   In my usage, a behavioral ensemble that has the effect of conserving transmission of some skill over time even when it is rarely the most advantageous among a set of possible alternatives. An example would be the ballistic capture of large, mobile prey. From a caloric perspective, this is almost never the most efficient use of time and energy in a foraging community—and yet, in most foraging communities it occupies significant time and energy among a large segment of the population. Under certain rare conditions of extrinsic risk (for instance, periodic drought), when the yields from other foraging strategies drop off precipitously, hunting large mobile prey and hoping for a windfall may be your best bet. But the skills needed to hunt effectively—tracking, ballistic conditioning—demand years of motor and observational learning. So in the absence of a behavioral ensemble that encourages their ongoing transmission, say, by valorizing ballistic heroics and the redistribution of meat, there's a risk these skills could be lost in between lean times.

Strategy   A covering term for any combination of behaviors by which an individual actor participates in a population game. Often, the behavior in question pertains to negotiating some kind of social encounter with other participants. So, for instance, males in some species of lizards found in western North America have been found to play a version of rock-paper-scissors, adopting different mating strategies (dominating a harem, mimicking females and mating clandestinely under the eyes of a dominant lizard, or cooperating for defense) that prevail in circuit fashion in the same way that paper prevails against rock, scissors against paper, and rock against scissors. Strategies in this sense need not be behavioral choices in the familiar sense—in the case of side-blotched lizards, a male's mating strategy is genetically determined. Evolutionary game theory models changes over time in the proportion of members of a population that adopt a particular strategy in this broad sense. Most versions of evolutionary game theory tend to make strong assumptions about the character of the revision protocols by which strategies change over time, and these may limit its applicability to certain phenomena commonly observed in human behavior (such as skills reservoirs). See also *game theory, landscape.*

Trait-transmission model   In this book, a model for the propagation of behavior that emphasizes the intergenerational transmission of discrete traits, possibly with allowance for a fixed rate of mutation. In the learning theory literature, commonly known as a replicator model. See chapters 1 and 2 for critique.

Treadmill   A metaphor for one commonly hypothesized threat to the cumulativity of human culture: the absence of a sufficient number of good learning models to support the intact transmission of adaptive traits. In this metaphor, social evolution unfolds on a demographic treadmill. It is rare to see the metaphor unpacked in any detail, but, roughly, you could imagine that the speed of the treadmill corresponds inversely to the size of the pool of good models for trait transmission (that is, a faster treadmill corresponds to a smaller effective population) while the rate at which the community runs on the treadmill corresponds to the fidelity with which it transmits adaptive traits over time (or corresponds inversely to the rate of maladaptive error in trait transmission). So at a given population size (1 / treadmill speed) a community must manage a given fidelity of transmission (1 / error rate) to maintain cumulativity.

Notes

vi *"He decides that after this . . ."*: Murnane 2019, 183.

PREFACE: LIVING EPIPHYTICALLY

xviii *to reintegrate anthropology*: Fuentes and Wiessner 2016.
xix *shunting*: Penny n.d.
xxi *an account of how anthropology developed as a discipline*: Kuklick 2011.
xxviii *actors' accounts of this transformation*: Brown 2009.
xxix *the stresses of displacement are more difficult to quantify*: Yu et al. 2020.
xxx *setting preventive fires practically daily*: Thomas Fuller, "Reducing Fire, and Cutting Carbon Emissions, the Aboriginal Way," *New York Times*, January 16, 2020, nyti.ms/2uO3Rpv.

1. TREADMILLS

2 *"'treadmill' of cultural loss"*: Kline and Boyd 2010.
3 *rates of innovation*: Bettencourt et al. 2007.

4   *In a 2001 paper*: Shennan 2001.

5   *first elaborated by population geneticist Ronald Fisher*: Fisher 1930. Fisher was founding chair of the Cambridge Eugenics Society. In 2020, Gonville and Caius College, Cambridge, responding to pressure from the Black Lives Matter and Rhodes Must Fall campaigns, removed a window honoring Fisher.

5   *enhanced cognitive capacities*: Wynn and Coolidge 2010. Cf. Berson 2019, chapter 3.

5   *In the years since evolutionary anthropologist Joseph Henrich*: Henrich 2004.

7   *Humans first entered Sahul*: Malaspinas et al. 2016; Tobler et al. 2017; Clarkson et al. 2017.

7   *Humans first entered Tasmania*: Cosgrove 1999; Colhoun and Shimeld 2012.

7   *optical luminescence dates*: Rhodes 2011. Optically stimulated luminescence (OSL) is a method of dating lithics (stone artifacts) and sedimentary depositional contexts. OSL relies on the fact that, over time, exposure to radiation from the decay of trace radionuclides in the soil or sediment induces electrons in the crystal lattice of quartz and other silicate minerals to enter an unstable excited (higher-energy, conduction-band) state. At defects in the lattice, a proportion of these electrons become trapped in states near the conduction band. These electrons remain in an excited state until they are freed by exposure to heat or light. When the electrons return to the ground state they emit light in characteristic frequency bands corresponding to the degree of excitation. Typically, one to two minutes of exposure to bright sunlight is sufficient to "bleach" a sample (return its electrons to the ground state). So if you can estimate background radiation for the period since deposition, the quantity of light emitted by a sample on stimulation with heat (thermoluminescence) or light (OSL) provides a measure of the time since the sample was deposited. Quartz OSL dates are precise and reliable for time depths of .001–200 ka (beyond 200 ka, electron trap saturation limits resolution, though in some cases the technique can be extended as far back as 1000 ka). OSL offers the additional advantage that dates can be obtained in the absence of the organic materials required for carbon–14 dating.

8   *continued occupation of the Bass Strait islands*: Bowdler 2015; Jones 1977.

9   *at the Last Glacial Maximum*: The timing and tempo of glaciation and deglaciation varied from region to region, with peri-Holocene warming trends in the South Pacific possibly preceding those in the North Atlantic. At the same time, new evidence suggests that the late Pleistocene witnessed a stronger coupling of decadal-to-millennial climate fluctuation between the tropical and polar Pacific than exists at present (i.e., a stronger El Niño Southern Oscillation effect). See Jones et al. 2018; Hall et al. 2010.

9   *keeping the land open using fire*: Colhoun and Shimeld 2012; Macphail 2010; Mariani et al. 2017; Mackenzie and Moss 2017; Fletcher et al. 2014.

9   *"It is not uncommon"*: Cosgrove 1999, 366. For the rest of this section, Cosgrove 1999; Jones 1995; Pike-Tay and Cosgrove 2002.

12   *To a remarkable degree*: Taylor 2014; Jones 1995.

12   *Jones was not alone*: See for example Diamond 1977.

13   *The first recorded visit*: Boyce 2008; Reynolds 2012; Johnson and McFarlane 2015.

14   *a reserve was established at Wybalenna*: Johnson and McFarlane 2015. In 1847, those remaining at Wybalenna, together with those returned from Port Phillip, were relocated to Oyster Cove, a disused penal settlement in the south that had been judged too unhealthy for convicts. Some of the women interned there escaped by marrying freed convicts. One couple was granted a plot of land outside the reserve and encouraged to make a go as homesteaders; white resentment frustrated their efforts. Some of the men left the reserve to become whalers. By 1855, there were five men and eleven women left at Oyster Cove. Upper respiratory disease and alcoholism picked them off, the last, known in the historical record as Truganini, dying in 1876. The present-day Tasmanian Aboriginal community is descended from those who managed to marry out of the reserve.

14   *Jones inferred*: Jones 1977, 321; Taylor 2014, 337.

15   *from what we know of the rest of Australia*: Berson 2014. For a recent effort at linguistic phylogeny, Bowern 2012. Bowern's analysis is limited by its reliance on word-lists, since grammatical structure (phonology, morphosyntax) may ramify and converge in ways orthogonal to lexicon. On top of this, Bayesian network phylogenies often take for granted a strict separation between

first- and second-language acquisition (transmission versus diffusion) that is problematic for precontact Australia.

15 *a population under siege*: Ryan 2013 estimates a precontact population of some 6,000, reduced, by 1831, to 250.

15 *accusations of willfully denying the continuance of the Aboriginal Tasmanian community*: Taylor 2014. Jones was a committed anticolonialist but found himself out of step with a new determination among Indigenous activists, starting in the 1970s, to reject the salvage mindset of an earlier generation of anthropologists.

16 *Plant-source foods . . . in the diet of mainland Australians*: Berson 2019, chap 3.

17 *"Reliable items in the diet of the Tasmanians were supplied by the women"*: Hiatt 1968, 217. On gendered differences in foraging strategy, Berson 2019, chapter 3; Bliege Bird and Bird 2008.

17 *a seal-hunting technique*: Hiatt 1968, 207–8. Hiatt speculates that the women's technique may have been influenced by the European sealers then in the area, who often employed Tasmanian women as seal hunters.

17 *there are times when a windfall strategy makes sense*: Jones, Bliege Bird, and Bird 2013.

18 *skills reservoir*: From the perspective of population game theory, a skills reservoir represents something like a strictly dominated strategy, a strategy that survives indefinitely despite the fact that it is never the best option for any participant (a skills reservoir could be a strategy that becomes a best option transiently, under a rare confluence of extrinsic circumstances). On the survival of strictly dominated strategies, Hofbauer and Sandholm 2011.

18 *where projectile weapons make sense*: Read 2006.

20 *Like so much of the evidence we've encountered thus far*: For the next two paragraphs, see Gilligan 2008.

20 *The archaeology of clothing presents unique difficulties*: Gilligan 2010.

22 *In Tasmania, the evidence favors a thermal hypothesis*: Gilligan 2007a, 2007b.

23 *Physiological adaptation to cold*: Snodgrass et al. 2007; Lidell and Enerbäck 2010; on the epigenetics of stress response, Mulligan 2016. In newborn humans, brown adipose tissue (BAT) is concentrated in the head, neck, shoulder blades, and axillae. Until recently, it was thought that humans did not retain significant BAT

deposits into adulthood. Newer noninvasive imaging techniques have identified a distinctive brown adipose tissue signature in the neck and upper back, with enhanced metabolic activity under cold stimulation. There is some evidence that long-term recurring cold exposure stimulates the growth of BAT deposits. Morphological adaptations (compact body, shorter limbs) may be developmental in origin, but over time individuals who are genetically predisposed to respond, in childhood and adolescence, to enhanced basal metabolic rate by growing in this way will be more likely to flourish as adults. It is also likely that morphological responses to enhanced metabolic rate in development are mediated in part by epigenetic signals, triggered in turn by enhanced metabolism and other components of a stress response to cold. These epigenetic signals may be transmitted with the genome in reproduction (see chapter 4). In these ways, developmental adaptations to the environment (e.g., enhanced metabolic rate in response to cold) can shape transgenerational adaptations such as changes in body plan.

25   *The ethnographic record of Tasmanians' use of fire*: Hiatt 1968, 211ff.

25   *"Fire-making was difficult in the damp Tasmanian climate . . ."*: Gott 2002, 654.

25   *a couple of points*: For these paragraphs, see Chazan 2017; Sandgathe 2017; Goldberg, Miller, and Mentzer 2017.

26   *time depths on the order of 400–800 ka*: Roebroeks and Villa 2011; Dibble et al. 2017.

28   *"On one of the last days of December 1947 . . ."*: Murnane 2019, 13.

29   *cosmological analogism*: Lloyd and Sivin 2003.

2. SCAFFOLDS

33   *a community's capacity to maintain an economic repertoire of a given complexity*: Collard et al. 2016.

35   *increased difficulty getting plant-source carbohydrates*: Andersson and Read 2016, 267.

35   *dyadic (two bodies in coordination) dynamics*: Berson 2015, chapter 1.

36   *by the number of junctures by which those components are configured into the whole*: This method is rare in archaeology. For its use in architecture, see Hillier 2007.

37   *"operatory chain"*: Haidle 2010.

37   *"As the organism of a fungus cannot be understood"*: Haidle 2016, 16.

38   *Scaffolding is not a new concept in the philosophy of evolution*: Wimsatt 2014; Sterelny 2011.

38   *somatic niche*: Berson 2015.

40   *enregisterment*: Agha 2007.

42   *debates about the validity of treadmill models*: Collard et al. 2016; Vaesen et al. 2016.

42   *considerable laboratory research*: Tomasello 2008.

42   *you do not need an explicit theory of mind*: Thompson 2007; Stewart, Gapenne, and Di Paolo 2010; Frith and Frith 2012.

43   *findings on real-world model selection strategies have been diverse*: MacDonald 2007; Lancy 2010; Terashima and Hewlett 2018.

44   *Transmission scenarios make no provision for practice*: Collard et al. 2016; Vaesen et al. 2016.

44   *selective constraints on innovation vary from one domain of behavior to another*: You could argue that what I'm pointing to is not a difference in degree of selective constraint but in the number of degrees of freedom implicated in the pattern of behavior—the number of factors that vary independently. In the case of the sound system of a language, there are many degrees of freedom—constraints on syllable shape; presence or absence of vowels, plosives, and fricatives with different modes of oral and nasal articulation; stress and tone contour; et cetera. In the case of locomotor style there may be fewer degrees of freedom—perhaps degree of involvement of the shoulders, hips, knees, wrists, and ankles; pitch of the upper body; which aspect of the plantar surface of the foot makes contact with the ground first; et cetera. When you have more degrees of freedom to work with, the selective pressure on any one is less—it is like having more vertebrae with which to articulate one's body to a curved surface. So selection will appear more relaxed when in fact it is simply more thinly distributed across individual characters of behavior.

I would respond in two ways. First, one thing this discussion demonstrates is that decomposing behavior into its constituent parts is not trivial. Sometimes, useful "cuts" jump out at us, such as a distinction between vowels, fricatives, and plosives or between hip involvement and knee involvement. But a fine-grained analysis of the spectral peaks in some high-dimensional pattern of behavior—which components contribute most—is more demanding. Often such a spectral analysis is not possible at all, or if it is possible it depends on further prior choices—e.g., for locomotor kinematics, which joints to track, how many cameras to use, and so on. That is, the decomposition is not just epistemically but ontologically indeterminate. This should give us pause when we encounter heuristic rubrics of behavior complexity of the sort that litter the skill transmission literature. (On the spectral analysis of locomotion, see Ellamil et al. 2016.)

Second, though sometimes what looks like relaxed selection may be the redistribution of selection across a broader range of features, there is no question that selection itself varies in its degree of tension or relaxation. To take a trivial example: think of how the seasonal rhythms of plant and animal life cycles play a weaker role in constraining your own diet than they did those of precontact Tasmanians.

45   *topography to exert a more stringent degree of selection on locomotor style*: Ingold 2005, citing unpublished work from 1996 by the Japanese anthropologist Junzo Kawada.

45   *"[A] single episode for attempting to obtain a seal through its breathing hole . . ."*: Read 2008, 602.

46   *"Among Fuegians, crafting bone-tipped arrows involved a 14-step process . . ."*: Henrich 2004, 204.

48   *cold-water breath-hold diving*: On Japan and Korea, see Rahn and Yokoyama 1965; Lee, Park, and Kim 2017; Lindholm and Lundgren 2009. On transgenerational adaptations to breath-hold diving, see Ilardo et al. 2018.

52   *fluffy-headed conjecture*: The challenge posed by factors that contribute to flourishing but not to adaptation is not that they are more difficult to operationalize, but that they are not susceptible of partial ordering in the way we expect of cardinal parameters. That is, the challenge is order-theoretic.

53   *human life history is exceptional, among animals, for its elongated character*: Kuzawa and Bragg 2012; Schwartz 2012; Berson 2019, chapters 1–2.

53   *collaborative foraging and childrearing*: See the essays of Hrdy, Hawkes, and Rosenberg in Calcagno and Fuentes 2012.

3. EQUILIBRIA

56   *something uniquely resonant about fire*: Compare Pyne 2016.

57   *their purpose in making fire*: Berson 2019, chapter 3.

58   *"It cannot be denied . . .":* Bunbury 1930, 105. Quoted in Hallam 1975, 47.

59   *like a marble balanced on a saddle that divides two basins*: Friedman and Sinervo 2016, 17.

59   *the recent climate history of what is now Tasmania*: For this passage, Wanner et al. 2008.

61   *"the most important branch of internal variability of the global climate system"*: Wanner et al. 2008, 1808.

61   *The vegetative regime of Southwest Australia*: Dodson 2001; Blome et al. 2012. For background on using the plant fossil record to estimate trends in climate, Willis and McElwain 2014.

62   *atmospheric water potential*: In Australia, high seasonal and interannual variance in rainfall makes mean precipitation a poor predictor of vegetative regime. Even if, over the long term, a region sees the kind of rain capable of supporting a mesic flora, its plants must still be capable of surviving extended dry spells.

63   *curation of surface water, herbivorous fauna, and extensive networks of cleared tracks*: Hallam 1975, 67–71.

63   *What about the habits of the humans*: Dortch, Balme, and Ogilvie 2012; Dortch and Wright 2010; Faith et al. 2017.

65   *constellation of changes in economic strategy and social structure*: Berson 2019, chapter 4.

67   *"Fire-Stick Farming"*: Jones 1969. For contemporary discussion, Sandgathe and Berna 2017; Scott et al. 2016.

68   *This awareness may or may not be linked to ascription of intent*: Bliege Bird and Bird 2008.

69   *pain asymbolia*: Craig 2009.

72   *peripersonal space*: Bufacchi and Iannetti 2018.

73   *"The process of burning . . .":* Noble 2006, 233.

73   *origins of controlled anthropogenic combustion*: Gowlett 2016; Goldberg, Miller, and Mentzer 2017; Mallol and Henry 2017; Roebroeks and Villa 2011; Dibble et al. 2017.

75   *negotiating a rugged terrain*: In the thought experiment that follows in the main text, you could think of the individual rows of the plot as the dynamic landscape of an evolutionary game at one

stage in its history. Extrinsic forcings modify that landscape over time, as does one's own activity. So it is not enough to identify a basin of attraction—a stable point in the landscape—and a Lyapunov trajectory that remains suitably within that basin. One must continually renegotiate the landscape as it changes.

4. LANDSCAPES

78   *"When he woke again it was still dark . . ."*: McCarthy 2006, 50.

79   *to turn* environment *into the more specific* landscape: Stilgoe 2016.

80   *the space of options is itself plastic*: Erwin 2017, discussed below. One could ask whether there is a homeomorphism (i.e., two-way isomorphic mapping) between two kinds of feature spaces, call them *Plastic* and *Static*. Then, Plastic would be defined in terms of well-defined features, where the space itself is mutable and evolution incorporates a process of niche construction in the feature space. In Static, by contrast, the transformations over time in Plastic would be represented simply as different regions of feature space. This has a certain appeal. It would allow us to model changes in the shape of Plastic as part of a search through an isomorphic space, Static, so we would not need to imagine two concurrent processes of evolution in Plastic, search through the existing space and modification of the space itself. If it existed, the "features" of Static would not be the same as the features of Plastic, or at least, Static would have to incorporate additional dimensions that did not correspond to those of adaptive spaces as we typically conceive them—dimensions for the plasticity of feature space conventionally understood. So perhaps the ontological unification would be illusory.

82   *if you imagine a space of two features*: In fact, it is possible to map a space of n features into a diagram of dimension $n - 1$ using what are called barycentric coordinates (equivalently, a ternary plot or de Finetti diagram). See Friedman and Sinervo 2016, 11, 52.

82   *"The folding of small RNA molecules . . ."*: Erwin 2017, 4, my emphasis.

83   *It is rather that they form a dense network with selective cues flowing bidirectionally between all pairs*: A dense network of topologies suggests a category-theoretic view of evolution. Imagine a

category *Evo*. Here, the different dimensions of variation (genetic, epigenetic, behavioral, environmental) represent the objects of Evo, while the flows of selective cues between them represent the arrows. These arrows need not be isomorphisms, as the divergence of topology between RNA sequence and three-dimensional conformation suggests.

83 *"Since roughly the same number of genes..."*: Erwin 2017, 4, my emphasis.

84 *"Molecular clock studies indicate..."*: Erwin 2017, 5, my emphasis. On polyploidy, Van der Peer, Mizrachi, and Marchal 2015.

85 *"the largest environmental cause of disease and premature death"*: Landrigan et al. 2017, 462.

86 *non-monotonic dose response*: Jacobs et al. 2017, 114.

87 *developmental reprogramming*: Jacobs et al. 2017, 115; cf. Mulligan 2016.

87 *permanent threshold shift ... depressed synaptic connectivity*: Gourévitch et al. 2014.

88 *impaired cardiovascular function and heightened risk of myocardial infarction and stroke*: Basner et al. 2014.

88 *Across a broad range of animal taxa*: Kight and Swaddle 2011.

88 *These effects appear to be heritable*: Miller and Raison 2016; Mulligan 2016.

88 *the most surprising risk associated with urban living is schizophrenia*: Lederbogen et al. 2011; Stepniak et al. 2014. The DAMP cascade represents an evolutionary response to the fact that tissue trauma is often preceded by social stress in the form of threatening behavior from a conspecific. See Fleshner, Frank, and Maier 2017.

89 *Oxidative stress*: For what follows, Isaksson 2015. Along with oxidative stress, Isaksson considers inflammation as a pathway linking environmental stress to impaired coping. Inflammation, of course, is a central component of immune response. It entails the targeted release of reactive oxygen and nitrogen species.

90 *The challenge of the urban environment is multimodal*: Halfwerk and Slabbekoorn 2015.

91 *for our capacity to feed ourselves*: The question is one not just of agricultural land and risk of extreme weather events but of the nutritional quality of food produced under historically unprecedented atmospheric CO_2 concentrations. See Smith and Myers

2018. Here again I ask the reader's generosity in allowing me this elastic use of the first-person plural—*we, our*. My intent is not to paper over disparities in access to food or living space, but rather to forbear on the nature of the structuring phenomena by which a population gives rise to a community or society.

92   *In a pair of elegant studies*: Obradovich and Fowler 2017; Obradovich et al. 2017.

93   *As I've discussed in earlier work*: Berson 2015.

94   *contact tracing during the COVID-19 pandemic*: COVID-19 National Emergency Response Center 2020; Baker et al. 2020.

95   *"worldwide trends in insufficient physical activity"*: Guthold et al. 2018. This study, in contrast to Obradovich and Fowler 2017, relied exclusively on surveys that "explicitly included physical activity across four key domains—i.e., for work, in the household (paid or unpaid), for transport to get to and from places (i.e., walking and cycling), and during leisure time (i.e., sports and active recreation)" (e1078).

96   *more than three times as great in low-income communities*: Obradovich et al. 2017. For reasons unexplained, the authors use a reported annual income of fifty thousand dollars as the cut-off for their "lower-income" category rather than keying the cut-off to income distribution quantiles, cost of living, or other measures of relative or absolute poverty.

96   *continued erosion of sleep*: On the ecology of sleep, Worthman 2011.

97   *one recent hypothesis*: Samson and Nunn 2015; Samson et al. 2017.

97   *Under laboratory conditions*: Wehr 2001. On early modern sleep, Ekirch 2005.

99   *Ultrasociality*: Gowdy and Krall 2016.

99   *anthropogenic biomes or anthromes*: Ellis 2015.

100   *the history of sociobiology and its successors*: See Berson 2014.

100   *In a careful review*: Powers and Lehmann 2017.

102   *"it is hard to find actual empirical cases . . ."*: Powers and Lehmann 2017, 905.

102   *the forms of classificatory kinship found in Australia*: Berson 2014; Berson 2019, chapter 3.

102   *"marked interpersonal differences"*: Powers and Lehmann 2017, 915.

102 *debate between proponents of inclusive fitness and those of group selection*: Leigh 2010. See also Lehmann et al. 2007.

104 *inductive generalization from that evidence to a hypothesized universal of human cognition*: Compare the discussion of the "Expensive Tissue Hypothesis" in Berson 2019, chapter 2.

104 *"The term* relationship *is a heuristic tool . . . In blood feuds . . .":* De Ruiter, Weston, and Lyon 2011, 561, 562.

105 *not what matter for formulating generalizations about social complexity*: De Ruiter, Weston, and Lyon 2011, 562–63. Further, neocortex size "would appear to be a rather poor predictor of group size, beyond a certain type of social relationship that represents an upscaled pair bond within culturally situated human populations."

105 *great emphasis on institutions*: Powers and Lehmann 2017, 915–16; cf. Powers and Lehmann 2013; Powers, Van Schaik, and Lehmann 2016. Powers and Lehmann cite Hurwicz 1996 for their formal characterization of institutions and Ostrom 1990 for an account of common-pool forest governance in the vicinity of Yamanakako-mura, near Mount Fuji.

106 *all the way down the discourse chain*: Agha 2007. On the iterative evolution of signaling games, compare Skyrms 2010.

107 *a curiously hypermasculine tang*: See for example, Rusch's account of parochial altruism, which attempts to explain the simultaneous appearance of cooperativeness and violent tendencies in humans: "The conventional contest mechanism used in the existing models randomly matches groups pairwise and compares them by the amount of in-group welfare they produce . . . . The group with lower welfare [by implication, the group that is less cooperative internally] is then replaced with a newly created group consisting of fitness proportionally procreated offspring of the individuals of the superior group. . . . More elaborate models allow for a calibration of the fraction of individuals killed in the defeated group, i.e. the 'brutality' of the victors. The majority of the models, however, simply assume the total annihilation of the losing group" (Rusch 2014, 2).

107 *a reaction-diffusion process*: Munafo n.d. The Gray-Scott model consists of a pair of differential equations formulated to describe the behavior of a pair of generic chemical reagents, conventionally known as U and V. U and V react to produce V (that is, V is

autocatalytic); *V* spontaneously converts into *P*, an inert product. The dynamic was first described by Alan Turing for the formation of pigmentation patterns in the scales of fish and reptiles (Turing 1952).

4BORO. LANDSCAPES AND SCAFFOLDS

112   The introduction of the landscape metaphor into evolutionary theory: Wright 1932; Fisher 1930. For context, Erwin 2017.

112   semantic templates: Levinson and Burenhult 2009.

113   image schemata: Johnson 2008.

118   Category theorists refer to their work, approvingly, as "abstract nonsense": McLarty 1990.

121   the "view from nowhere": Nagel 1986.

5. DITCH KIT

128   A Million Random Digits with 100,000 Normal Deviates: Nahum et al. 2017.

128   the gluon, to use philosopher Graham Priest's term: Priest 2014.

129   sleep in their cars at night because they cannot afford housing: N. Buhayar and E. Deprez, "The Homeless Crisis Is Getting Worse in America's Richest Cities," Bloomberg Businessweek, November 20, 2018.

129   the number of people in all parts of the world who were "forcibly displaced": UNHCR 2017.

129   The Life-Changing Magic of Tidying Up: Kondo 2014.

130   much of the energy behind the decluttering movement: Sasaki 2017; Matsumoto 2018; Magnusson 2017.

130   as a digital nomad, or in one of a proliferating number of "dormitories for grown-ups": K. Chayka, "When You're a 'Digital Nomad,' the World Is Your Office," New York Times, February 8, 2018, nyti.ms/2BMKoLC; N. Bowles, "Dorm Living for Professionals Comes to San Francisco," New York Times, March 4, 2018, nyti.ms/2D2dymk.

132   "the sound of an oar hitting waves . . . nothing to fear on the roads": Reichhold 2008, 54, 58, 71.

133   a tension that is immanent in settled life: This is a theme that a number of writers have struggled to formulate clearly. Scott

(2009) relies extensively on Clastres's argument ([1974] 1989) that the Tupi-Guarani represent not a relict society that failed to develop sedentism but one that had expressly rejected it. Compare the closing remarks in Berson 2014 on the relationships between genealogy, property, and society.

134   *"the most symbolically dense shells are known as 'chiefly shells'..."*: Weiner 1994, 395–96. Compare Munn [1986] 1992.

134   yams *"are used to challenge adversaries or authenticate status"*: Weiner 1994, 392.

135   *"so dense with cultural meaning..."*: Weiner 1994, 394.

135   *the way works of art circulate in the auction markets*: Compare Chatwin 1987.

136   *multiple bookkeeping, reserving certain caches of value for certain kinds of transactions*: Guyer 2004; Hutchinson 1996; Zelizer 2007. Hutchinson notes how her Nuer interlocutors of the 1980s distinguished six registers of capital, ranging from "the cattle of daughters" (i.e., cattle brought into a household as bridewealth) to "the cattle of money" (cattle purchased with cash), and "the money of work" (cash earned by the sale of agricultural and hunting surpluses), down to "the money of shit" (that earned by low-status wage labor).

136   *"He begins as a small boy in grade two..."*: Murnane 2019, 152–55, my emphases.

138   *the skillful management of stuff*: Something similar could be said about time. The relationship between stuff-affluence and time-poverty is an area of active research, as the role of consumers in household carbon footprint mitigation seems to entail a tradeoff between discretionary time and energy use. Despite forty years of effort, widely adopted conventions for the operationalization of time poverty continue to elude economists and sociologists. See Williams, Masuda, and Tallis 2016; Wiedenhofer et al. 2018.

139   *the Atlantic slave trade*: Green 2019.

140   *an increasingly meronomic—concerned with part-whole relationships—view of our bodies*: Hacking 2007.

142   *Great Pacific Garbage Patch*: Lebreton et al. 2018; cf. Cózar et al. 2017. On "missing plastic" below the ocean surface, Worm et al. 2017, 8.

143   *"Production, use, and fate of all plastics ever made"*: Geyer, Jambeck, and Lavender Law 2017.

143 *China announced new restrictions*: Brooks, Wang, and Jambeck 2018.

143 *plastic debris seems to serve as a vector for organic and heavy-metal pollutants*: Worm et al. 2017, 13–15.

144 *taxonomic composition of the Earth's biomass*: Bar-On, Phillips, and Milo 2018.

145 *"trophic downgrading"*: Estes et al. 2011.

146 *fire here represents an "abiotic herbivore"*: Malhi et al. 2016; Bakker et al. 2016.

146 *zoonotic infections . . . dense coresidentiality*: Johnson et al. 2020. Chiroptera—bats—are overrepresented as reservoirs, along with rodents, primates, and livestock. SARS-CoV-2 is thought to have originated in a bat population.

146 *the incidence and magnitude of these events are increasing*: Fey et al. 2015.

147 *the rising incidence of coastal flooding . . . decline of polar sea-ice cover*: Vitousek et al. 2017; Coumou et al. 2018.

147 *it will be more resistant to perturbance*: Steffen et al. 2018. Reviews of the type I have offered run the risk of inductive circularity: If our basis for attributing regime shift to the Earth system comes in part from observations of the frequency and magnitude of extreme events, how can we then attribute extreme events of the same kinds to that regime shift? This is a deep question, but I can at least describe the methods that have been accepted in one field that deals regularly with the attribution of extreme events: meteorology. The basic strategy is to assume that the return process for extreme weather events of a given magnitude will follow a Poisson distribution—that is, assume that the frequency with which events of a given magnitude "arrive" or "return" varies about a mean in the manner expected if the mean interval between events is independent of how long you've been "waiting." (Poisson distributions are common in the natural and social worlds and have applications to the modeling of a range of arrival phenomena, including those implicated not only in weather but also in disease incidence, traffic congestion, etc. The Poisson distribution is distinguished by the fact that the shape of a particular instance of the distribution is governed by just one term, the mean rate of arrival or incidence.) Take a historical period for which you have data about the frequency of events of a particular type—heat waves, wildfires,

typhoons—of different magnitudes. From these data, estimate a series of means for events across a range of magnitudes. Then, estimate a series of means for a "period of interest"—a period, perhaps overlapping with the baseline period, for which you'd like to estimate the degree to which some change in the climate regime has affected the incidence and magnitude of the type of event in question. You can think of these two series of Poisson means as curves on a plot. The difference between the two curves tells you what proportion of the events in the period of interest is attributable to (human) activity of a form stipulated to be absent from the baseline period. Whether you emphasize the difference in magnitude for events in the baseline period and period of interest with equivalent frequencies or the difference in frequency for events of equivalent magnitude will make a difference in the degree to which you hold human activity responsible. See Otto 2017.

147   *Corvids, cetaceans, and other-than-human primates*: Some readers may feel I have unfairly excluded cephalopods from this list. The reason has nothing to do with skepticism about the intelligence of octopuses, squids, and cuttlefish. Rather it concerns the nature of that intelligence, which, both in evolutionary and ontogenetic terms, seems to arise from selective pressures other than those that have shaped intelligence in the synapsids. If anything, the case of cephalopods shows us that social cognition is but one path among many toward the evolution of sapience. See Amodio et al. 2019.

148   *the strange way humans imitate others*: Hurley 2008.

149   *not on a treadmill but on a dance floor*: Tomasello 2008; Fitch 2015.

149   *The New Analog*: Krukowski 2017, chapter 3.

151   *frequency of extreme events*: *Extreme* as opposed to *extremal*: what is at issue here is not the kurtosis or tail-heaviness of a distribution of events but the proportion of those events that are extreme in the sense that they challenge our capacity to cope.

154   *enhanced weathering*: Lenton et al. 2012; Porada et al. 2016; Lawford-Smith and Currie 2017.

POSTSCRIPT: FOAMINESS

163   *decision theory as it was formulated*: Erickson et al. 2013.

163   *Graham Martin's stirring* The Sensory Ecology of Birds: Martin 2017.

163   *in line with my prior expectations about how enminded presences
       move*: Clark 2013.

166   *central pattern generation, the internal coordination of metabolic
       and respiratory processes, likely preceded sensorimotor coordi-
       nation*: Jékely, Keijzer, and Godfrey-Smith 2015; Godfrey-Smith
       2016; Azzalini, Rebollo, and Tallon-Baudry 2019.

167   *"Foams"*: For a start, see the essays on biological individuality col-
       lected in Pradeu 2016.

Sources

Agha, A. 2007. *Language and Social Relations*. Cambridge: Cambridge University Press.

Amodio, P., M. Boeckle, A. Schnell, L. Ostojic, G. Fiorito, and N. Clayton. 2019. "Grow Smart and Die Young: Why Did Cephalopods Evolve Intelligence?" *Trends in Ecology and Evolution* 34(1): 45–56.

Andersson, C., and D. Read. 2016. "The Evolution of Cultural Complexity." *Current Anthropology* 57(3): 261–86.

Azzalini, D., I. Rebollo, and C. Tallon-Baudry. 2019. "Visceral Signals Shape Brain Dynamics and Cognition." *Trends in Cognitive Sciences* 23(6): 488–509.

Baker, M., A. Kvalsvig, A. Verrall, L. Telfar-Barnard, and N. Wilson. 2020. "New Zealand's Elimination Strategy for COVID-19 and What Is Required to Make It Work." *New Zealand Medical Journal* 133(1512): 10–14.

Bakker, E., J. Gill, C. Johnson, F. Vera, C. Sandom, G. Asner, and J.-C. Svenning. 2016. "Combining Paleo-data and Modern Exclosure Experiments to Assess the Impact of Megafauna Extinctions on Woody Vegetation." *Proceedings of the National Academy of Sciences of the United States of America* 113(4): 847–55.

Bar-On, Y., R. Phillips, and R. Milo. 2018. "The Biomass Distribution on Earth." *Proceedings of the National Academy of Sciences of the United States of America* 115(25): 6506–11.

Basner, M., W. Babisch, A. Davis, M. Brink, C. Clark, S. Janssen, and S. Stansfeld. 2014. "Auditory and Non-auditory Effects of Noise on Health." *Lancet* 383(9925): 1325–32.

Becker, A., H. Nagel, C. Wolfrum, and I. Burger. 2016. "Anatomical Grading for Metabolic Activity of Brown Adipose Tissue." *PLOS ONE* 11(2): e0149458.

Berson, J. 2014. "The Dialectal Tribe and the Doctrine of Continuity." *Comparative Studies in Society and History* 56(2): 381–418.

———. 2015. *Computable Bodies: Instrumented Life and the Human Somatic Niche.* London: Bloomsbury.

———. 2019. *The Meat Question: Animals, Humans, and the Deep History of Food.* Cambridge, MA: MIT Press.

Bettencourt, L., J. Lobo, D. Helbing, C. Kühnert, and G. West. 2007. "Growth, Innovation, Scaling, and the Pace of Life in Cities." *Proceedings of the National Academy of Sciences of the United States of America* 104(17): 7301–6.

Bliege Bird, R., and D. Bird. 2008. "Why Women Hunt: Risk and Contemporary Foraging in a Western Desert Aboriginal Community." *Current Anthropology* 49(4): 655–93.

Blome, M., A. Cohen, C. Tryon, A. Brooks, and J. Russell. 2012. "The Environmental Context for the Origins of Modern Human Diversity: A Synthesis of Regional Variability in African Climate 150,000–30,000 Years Ago." *Journal of Human Evolution* 62(5): 563–92.

Bowdler, S. 2015. "The Bass Strait Islands Revisited." *Quaternary International* 385: 206–18.

Bowern, C. 2012. "The Riddle of Tasmanian Languages." *Proceedings of the Royal Society B* 279(1747): 4590–95.

Boyce, J. 2008. *Van Diemen's Land.* Melbourne: Black Inc.

Brooks, A., S. Wang, and J. Jambeck. 2018. "The Chinese Import Ban and Its Impact on Global Plastic Waste Trade." *Science Advances* 4(6): eaat0131.

Brown, T. 2009. *Change by Design: How Design Thinking Transforms Organizations and Inspires Innovation.* New York: HarperCollins.

Bufacchi, R., and G. Iannetti. 2018. "An Action Field Theory of Peripersonal Space." *Trends in Cognitive Sciences* 22(12): 1076–90.

Bunbury, H. 1930. *Early Days in Western Australia: Being the Letters and Journals of Lieutenant H. W. Bunbury.* London: Oxford University Press.

Calcagno, J., and A. Fuentes, eds. 2012. "What Makes Us Human? Answers from Evolutionary Anthropology." Special issue, *Evolutionary Anthropology* 21(5): 182–94.

Chatwin, B. 1987. *The Songlines.* New York: Viking.

Chazan, M. 2017. "Toward a Long Prehistory of Fire." *Current Anthropology* 58(S16): S351–59.

Clark, A. 2013. "Whatever Next? Predictive Brains, Situated Agents, and the Future of Cognitive Science." *Behavioral and Brain Sciences* 36(3): 181–253.

Clarkson, C., Z. Jacobs, B. Marwick, R. Fullager, L. Wallis, et al. 2017. "Human Occupation of Northern Australia by 65,000 Years Ago." *Nature* 547(7663): 306–10.

Clastres, P. [1974] 1989. *Society against the State.* Translated by R. Hurley and A. Stein. New York: Zone.

Colhoun, E., and P. Shimeld. 2012. "Late-Quaternary Vegetation History of Tasmania from Pollen Records." In *Peopled Landscapes: Archaeological and Biogeographic Approaches to Landscapes*, edited by S. Haberle and B. David, 297–328. Canberra: Australian National University Press.

Collard, M., K. Vaesen, R. Cosgrove, and W. Roebroeks. 2016. "The Empirical Case against the 'Demographic Turn' in Paleolithic Archaeology." *Philosophical Transactions of the Royal Society B* 371(S1698): 20150242.

Cosgrove, R. 1999. "Forty-Two Degrees South: The Archaeology of Late Pleistocene Tasmania." *Journal of World Prehistory* 13(4): 357–402.

Coumou, D., G. Di Capua, S. Vavrus, L. Wang, and S. Wang. 2018. "The Influence of Arctic Amplification on Mid-Latitude Summer Circulation." *Nature Communications* 9: 2959.

Covid-19 National Emergency Response Center. 2020. "Contact Transmission of Covid-19 in South Korea: Novel Investigative Techniques for Tracing Contacts." *Osong Public Health and Research Perspectives* 11(1): 60–63.

Cózar, A., E. Martí, C. Duarte, J. García-de-Lomas, E. van Sebille, et al. 2017. "The Arctic Ocean as a Dead End for Floating Plastics in the North Atlantic Branch of the Thermohaline Circulation." *Science Advances* 3(4): e1600582.

Craig, A. 2009. "How Do You Feel—Now? The Anterior Insula and Human Awareness." *Nature Reviews Neuroscience* 10(1): 59–70.

De Ruiter, J., G. Weston, and S. Lyon. 2011. "Dunbar's Number: Group Size and Brain Physiology in Humans Reexamined." *American Anthropologist* 113(4): 557–68.

Diamond, J. 1977. "Colonization Cycles in Man and Beast." *World Archaeology* 8(3): 249–61.

Dibble, H., A. Abodolahzadeh, V. Aldeias, P. Goldberg, S. McPherron, and D. Sandgathe. 2017. "How Did Hominins Adapt to Ice Age Europe without Fire?" *Current Anthropology* 58(S16): S278–87.

Dodson, J. 2001. "Holocene Vegetation Change in the Mediterranean-Type Climate Regions of Australia." *The Holocene* 11(6): 673–80.

Dortch, J., J. Balme, and J. Ogilvie. 2012. "Aboriginal Response to Late Quaternary Environmental Change in a Mediterranean-Type Region: Zooarchaeological Evidence from Southwestern Australia." *Quaternary International* 264: 121–34.

Dortch, J., and R. Wright. 2010. "Identifying Palaeo-Environments and Changes in Aboriginal Subsistence from Dual-Patterned Faunal Assemblages, South-Western Australia." *Journal of Archaeological Science* 37(5): 1053–64.

Ekirch, A. 2005. *At Day's Close: Night in Times Past*. New York: Norton.

Ellamil, M., J. Berson, J. Wong, L. Buckley, and D. Margulies. 2016. "One in the Dance: Musical Correlates of Group Synchrony in a Real-World Club Environment." *PLOS ONE* 11(10): e0164783.

Ellis, E. 2015. "Ecology in an Anthropogenic Biosphere." *Ecological Monographs* 85(3): 287–331.

Erickson, P., J. Klein, L. Daston, R. Lemov, T. Sturm, and M. Gordin. 2013. *How Reason Almost Lost Its Mind: The Strange Career of Cold War Rationality*. Chicago: University of Chicago Press.

Erwin, D. 2017. "The Topology of Evolutionary Novelty and Innovation in Macroevolution." *Philosophical Transactions of the Royal Society B* 372(1735): 20160422.

Estes, J., J. Terborgh, J. Brashares, M. Power, J. Berger, et al. 2011. "Trophic Downgrading of Planet Earth." *Science* 333(6040): 301–6.

Faith, J., J. Dortch, C. Jones, J. Shulmeister, and K. Travouillon. 2017. "Large Mammal Species Richness and Late Quaternary Precipitation Change in South-Western Australia." *Journal of Quaternary Science* 32(6): 760–69.

Fey, S., A. Siepielski, S. Nusslé, K. Cervantes-Yoshida, J. Hwan, et al. 2015. "Recent Shifts in the Occurrence, Cause, and Magnitude of Animal Mass Mortality Events." *Proceedings of the National Academy of Sciences of the United States of America* 112(4): 1083–88.

Fisher, R. 1930. *The Genetical Theory of Natural Selection.* Oxford: Clarendon.

Fitch, W. 2015. "Four Principles of Bio-Musicology." *Philosophical Transactions of the Royal Society B* 370(1664): 20140091.

Fleshner, M., M. Frank, and S. Maier. 2017. "Danger Signals and Inflammasomes: Stress-Evoked Sterile Inflammation in Mood Disorders." *Neuropsychopharmacology Reviews* 42: 36–45.

Fletcher, M.-S., B. Wolfe, C. Whitlock, D. Pompeani, H. Heijnis, et al. 2014. "The Legacy of Mid-Holocene Fire on a Tasmanian Montane Landscape." *Journal of Biogeography* 41(3): 476–88.

Friedman, D., and B. Sinervo. 2016. *Evolutionary Games in Natural, Social, and Virtual Worlds.* New York: Oxford University Press.

Frith, C., and U. Frith. 2012. "Mechanisms of Social Cognition." *Annual Review of Psychology* 63: 287–313.

Fuentes, A., and P. Wiessner, eds. 2016. "Reintegrating Anthropology: From Inside Out." Supplement, *Current Anthropology* 57 (S13).

Geyer, R., J. Jambeck, and K. Lavender Law. 2017. "Production, Use, and Fate of All Plastics Ever Made." *Science Advances* 3(7): e1700782.

Gilligan, I. 2007a. "Clothing and Modern Human Behaviour: Prehistoric Tasmania as a Case Study." *Archaeology in Oceania* 42(3): 102–11.

———. 2007b. "Resisting the Cold in Ice Age Tasmania: Thermal Environment and Settlement Strategies." *Antiquity* 81(313): 555–68.

———. 2008. "Clothing and Climate in Aboriginal Australia." *Current Anthropology* 49(3): 487–95.

———. 2010. "The Prehistoric Development of Clothing: Archaeological Implications of a Thermal Model." *Journal of Archaeological Method and Theory* 17(1): 15–80.

Godfrey-Smith, P. 2016. *Other Minds: The Octopus, the Sea, and the Deep Origins of Consciousness.* New York: Farrar, Straus and Giroux.

Goldberg, P., C. Miller, and S. Mentzer. 2017. "Recognizing Fire in the Paleolithic Archaeological Record." *Current Anthropology* 58(S16): S175–90.

Gott, B. 2002. "Fire-Making in Tasmania Absence of Evidence Is Not Evidence of Absence." *Current Anthropology* 43(4): 650–56.

Gourévitch, B., J.-M. Edeline, F. Occelli, and J. Eggermont. 2014. "Is the Din Really Harmless? Long-Term Effects of Non-traumatic Noise on the Adult Auditory System." *Nature Reviews Neuroscience* 15(7): 483–91.

Gowdy, J., and L. Krall. 2016. "The Economic Origins of Ultrasociality." *Behavioral and Brain Sciences* 39: e92.

Gowlett, J. 2016. "The Discovery of Fire by Humans: A Long and Convoluted Process." *Philosophical Transactions of the Royal Society B* 371(1696): 20150164.

Green, T. 2019. *A Fistful of Shells: West Africa from the Rise of the Slave Trade to the Age of Revolution*. Chicago: University of Chicago Press.

Guthold, R., G. Stevens, L. Riley, and F. Bull. 2018. "Worldwide Trends in Insufficient Physical Activity from 2001 to 2016: A Pooled Analysis of 358 Population-Based Surveys with 1.9 Million Participants." *Lancet Global Health* 6(10): e1077–86.

Guyer, J. 2004. *Marginal Gains: Monetary Transactions in Atlantic Africa*. Chicago: University of Chicago Press.

Hacking, I. 2007. "Our Neo-Cartesian Bodies in Parts." *Critical Inquiry* 34(1): 78–105.

Haidle, M. 2010. "Working-Memory Capacity and the Evolution of Modern Cognitive Potential: Implications from Animal and Early Human Tool Use." *Current Anthropology* 51(S1): S149–66.

———. 2016. "Lessons from Tasmania: Cultural Performance versus Cultural Capacity." In *The Nature of Culture*, edited by M. Haidle, N. Conard, and M. Bolus, 7–17. Dordrecht: Springer.

Halfwerk, W., and H. Slabbekoorn. 2015. "Pollution Going Multimodal: The Complex Impact of the Human-Altered Sensory Environment on Animal Perception and Performance." *Biology Letters* 11(4): 20141051.

Hall, B., G. Denton, A. Fountain, C. Hendy, and G. Henderson. 2010. "Antarctic Lakes Suggest Millennial Reorganizations of Southern Hemisphere Atmospheric and Oceanic Circulation." *Proceedings of the National Academy of Sciences of the United States of America* 107(50): 21355–59.

Hallam, S. 1975. *Fire and Hearth: A Study of Aboriginal Usage and European Usurpation in South-Western Australia*. Canberra: Australian Institute of Aboriginal Studies.

Henrich, J. 2004. "Demography and Cultural Evolution: How Adaptive Cultural Processes Can Produce Maladaptive Losses: The Tasmanian Case." *American Antiquity* 69(2): 197–215.

Hiatt, B. 1967, 1968. "The Food Quest and the Economy of the Tasmanian Aborigines." Pts. 1 and 2. *Oceania* 38(2): 99–133; 38(3)190–219.

Hillier, B. 2007. *Space Is the Machine: A Configurational Theory of Architecture*. London: SpaceSyntax.

Hofbauer, J., and W. Sandholm. 2011. "Survival of Dominated Strategies under Evolutionary Dynamics." *Theoretical Economics* 6(3): 341–77.

Hurley, S. 2008. "The Shared Circuits Model (SCM): How Control, Mirroring, and Simulation Can Enable Imitation, Deliberation, and Mindreading." *Behavioral and Brain Sciences* 31(1): 1–58.

Hurwicz, L. 1996. "Institutions as Families of Game Forms." *Japanese Economic Review* 47(2): 113–32.

Hutchinson, S. 1996. *Nuer Dilemmas: Coping with Money, War, and the State*. Berkeley: University of California Press.

Ilardo, M., I. Moltke, T. Korneliussen, J. Cheng, A. Stern, et al. 2018. "Physiological and Genetic Adaptations to Diving in Sea Nomads." *Cell* 173(3): 569–80.

Ingold, T. 2005. "Culture on the Ground: The World Perceived through the Feet." *Journal of Material Culture* 9(3): 315–40.

Isaksson, C. 2015. "Urbanization, Oxidative Stress, and Inflammation: A Question of Evolving, Acclimatizing, or Coping with Urban Environmental Stress." *Functional Ecology* 29(7): 913–23.

Jacobs, M., E. Marczylo, C. Guerrero-Bosagna, and J. Rüegg. 2017. "Marked for Life: Epigenetic Effects of Endocrine Disrupting Chemicals." *Annual Review of Environment and Resources* 42: 105–60.

Jékely, G., F. Keijzer, and P. Godfrey-Smith. 2015. "An Option Space for Early Neural Evolution." *Philosophical Transactions of the Royal Society B* 370(1684): 20150181.

Johnson, C., P. Hitchens, P. Pandit, J. Rushmore, T. Smiley Evans, C. Young, and M. Doyle. 2020. "Global Shifts in Mammalian Population Trends Reveal Key Predictors of Virus Spillover Risk." *Proceedings of the Royal Society B* 287(1924): 20192736.

Johnson, M. 2008. *The Meaning of the Body: Aesthetics of Human Understanding*. Chicago: University of Chicago Press.

Johnson, M., and I. McFarlane. 2015. *Van Diemen's Land: An Aboriginal History*. Sydney: University of New South Wales Press.

Jones, J., R. Bliege Bird, and D. Bird. 2013. "To Kill a Kangaroo: Understanding the Decision to Pursue High-Risk/High-Gain Resources." *Proceedings of the Royal Society B* 280(1767): 20131210.

Jones, R. 1969. "Fire-Stick Farming." *Australian Natural History* 16: 224–228.

———. 1977. "Man as an Element of a Continental Fauna." In *Sunda and Sahul: Prehistoric Studies in Southeast Asia, Melanesia and Australia*, edited by J. Allen, J. Golson, and R. Jones, 317–86. London: Academic Press.

———. 1995. "Tasmanian Archaeology." *Annual Review of Anthropology* 24: 423–46.

Jones, T., W.Roberts, E. Steig, K. Cuffey, B. Markle, and J. White. 2018. "Southern Hemisphere Climate Variability Forced by Northern Hemisphere Ice-Sheet Topography." *Nature* 554(7692): 351–55.

Kight, C., and J. Swaddle. 2011. "How and Why Environmental Noise Impacts Animals: An Integrative, Mechanistic Review." *Ecology Letters* 14(10): 1052–61.

Kline, M., and R. Boyd. 2010. "Population Size Predicts Technological Complexity in Oceania." *Proceedings of the Royal Society B* 277(1693): 2559–64.

Kondo, M. 2014. *The Life-Changing Magic of Tidying Up: The Japanese Art of Decluttering and Organizing*. Translated by C. Hirano, C. Berkeley: Ten Speed.

Krukowski, D. 2017. *The New Analog: Listening and Reconnecting in a Digital World*. New York: The New Press.

Kuklick, H. 2011. "Personal Equations: Reflections on the History of Fieldwork, with Special Reference to Sociocultural Anthropology." *Isis* 102(1): 1–33.

Kuzawa, C., and J. Bragg. 2012. "Plasticity in Human Life History Strategy: Implications for Contemporary Human Variation and the Evolution of Genus *Homo*." *Current Anthropology* 53(S6): S369–82.

Lancy, D. 2010. "Learning 'from Nobody': The Limited Role of Teaching in Folk Models of Children's Development." *Childhood in the Past* 3(1): 79–106.

Landrigan, P., R. Fuller, N. Acosta, O. Adeyi, R. Arnold, et al. 2017. "The *Lancet* Commission on Pollution and Health." *Lancet* 391(10119): 462–512.

Lawford-Smith, H., and A. Currie. 2017. "Accelerating the Carbon Cycle: The Ethics of Enhanced Weathering." *Biology Letters* 13(4): 20160859.

Lebreton, L., B. Slat, R. Ferrari, B. Sainte-Rose, J. Aitken, et al. 2018. "Evidence that the Great Pacific Garbage Patch Is Rapidly Accumulating Plastic." *Scientific Reports* 8(1): 4666.

Lederbogen, F., P. Kirsch, L. Haddad, F. Streit, H.Tost, et al. 2011. "City Living and Urban Upbringing Affect Neural Social Stress Processing in Humans." *Nature* 474(7352): 498–501.

Lee, J.-Y., J. Park, and S. Kim. 2017. "Cold Adaptation, Aging, and Korean Women Divers (*Haenyeo*)." *Journal of Physiological Anthropology* 36: Article 33.

Lehmann, L., L. Keller, S. West, and D. Roze. 2007. "Group Selection and Kin Selection: Two Concepts but One Process." *Proceedings of the National Academy of Sciences* 104(16): 6736–39.

Leigh, E. 2010. "The Group Selection Controversy." *Journal of Evolutionary Biology* 23(1): 6–19.

Lenton, T., M. Crouch, M. Johnson, N. Pires, and L. Dolan. 2012. "First Plants Cooled the Ordovician." *Nature Geoscience* 5(1): 86–89.

Levinson, S., and N. Burenhult. 2009. "Semplates: A New Concept in Lexical Semantics?" *Language* 85(1): 153–74.

Lidell, M., and S. Enerbäck. 2010. "Brown Adipose Tissue: A New Role in Humans?" *Nature Reviews Endocrinology* 6(6): 319–25.

Lindholm, P., and C. Lundgren. 2009. "The Physiology and Pathophysiology of Human Breath-Hold Diving." *Journal of Applied Physiology* 106(1): 284–92.

Lloyd, G., and N. Sivin. 2003. *The Way and the Word: Science and Medicine in Early China and Greece*. New Haven: Yale University Press.

MacDonald, K. 2007. "Cross-Cultural Comparison of Learning in Human Hunting." *Human Nature* 18(4): 386–402.

Mackenzie, L., and P. Moss. 2017. "A Late Quaternary Record of Vegetation and Climate Change from Hazards Lagoon, Eastern Tasmania." *Quaternary International* 432: 58–65.

Macphail, M. 2010. "The Burning Question: Claims and Counter Claims on the Origin and Extent of Buttongrass Moorland (Blanket Moor) in Southwest Tasmania during the Present Glacial-Interglacial." In *Altered Ecologies: Fire, Climate and Human Influence on Terrestrial Landscapes*, edited by S. Haberle, J. Stevenson, and M. Prebble, 323–40. Canberra: Australian National University Press.

Magnusson, M. 2017. *The Gentle Art of Swedish Death Cleaning: How to Free Yourself and Your Family from a Lifetime of Clutter*. New York: Scribner.

Malaspinas, A.-S., M. Westaway, C. Muller, V. Sousa, O. Lao, et al. 2016. "A Genomic History of Aboriginal Australia." *Nature* 538(7624): 207–14.

Malhi, Y., C. Doughty, M. Galetti, F. Smith, J.-C. Svenning, and J. Terborgh. 2016. "Megafauna and Ecosystem Function from the Pleistocene to the Anthropocene." *Proceedings of the National Academy of Sciences of the United States of America* 113(4): 838–46.

Mallol, C., and A. Henry. 2017. "Ethnoarchaeology of Paleolithic Fire: Methodological Considerations." *Current Anthropology* 58(S16): S217–29.

Mariani, M., S. Connor, M.-S. Fletcher, M. Theuerkauf, P. Kuneš, et al. 2017. "How Old Is the Tasmanian Cultural Landscape? A Test of Landscape Openness Using Quantitative Land-Cover Reconstructions." *Journal of Biogeography* 44(10): 2410–20.

Martin, G. 2017. *The Sensory Ecology of Birds*. Oxford: Oxford University Press.

Matsumoto, S. 2018. *A Monk's Guide to a Clean House and Mind*. Translated by I. Samhammer, I. London: Penguin.

McCarthy, C. 2006. *The Road*. New York: Knopf.

McLarty, C. 1990. "The Use and Abuse of the History of Topos Theory." *British Journal for the Philosophy of Science* 41(3): 351–75.

Miller, A., and C. Raison. 2016. "The Role of Inflammation in Depression: From Evolutionary Imperative to Modern Treatment Target." *Nature Reviews Immunology* 16(1): 22–34.

Mulligan, C. 2016. "Early Environments, Stress, and the Epigenetics of Human Health." *Annual Review of Anthropology* 45: 233–49.

Munafo, R. n.d. [2014]. "Stable Localized Moving Patterns in the 2-D Gray-Scott Model." arxiv.org/abs/1501.01990.

Munn, N. [1986] 1992. *The Fame of Gawa: A Symbolic Study of Value Transformation in a Massim Society*. Durham, NC: Duke University Press.

Murnane, G. [1974] 2019. *Tamarisk Row*. Sheffield: And Other Stories.

Nagel, T. 1986. *The View from Nowhere*. New York: Oxford University Press.

Nahum, J. Berson, S. Bullock, and H. Shomar. 2017. *A Million Random Digits*. Performed at the Haus der Kulturen der Welt, Berlin, December 2. Video of live performance, 41:48. hkw.de/de/app/mediathek/video/62060.

Noble, G. 2006. *Neolithic Scotland: Timber, Stone, Earth and Fire*. Edinburgh: Edinburgh University Press.

Obradovich, N., and J. Fowler. 2017. "Climate Change May Alter Human Physical Activity Patterns." *Nature Human Behavior* 1: 0097.

Obradovich, N., R. Migliorini, S. Mednick, and J. Fowler. 2017. "Nighttime Temperature and Human Sleep Loss in a Changing Climate." *Science Advances* 3(5): e1601555.

Ostrom, E. 1990. *Governing the Commons: The Evolution of Institutions for Collective Action*. Cambridge: Cambridge University Press.

Otto, F. 2017. "Attribution of Weather and Climate Events." *Annual Review of Environment and Resources* 42: 627–646.

Penny, S. n.d. "Orthogonal." Accessed January 2020. simonpenny.net /orthogonal.

Pike-Tay, A., and R. Cosgrove. 2002. "From Reindeer to Wallaby: Recovering Patterns of Seasonality, Mobility, and Prey Selection in the Palaeolithic Old World." *Journal of Archaeological Theory and Method* 9(1): 101–46.

Porada, P., T. Lenton, A. Pohl, B. Weber, L. Mander, et al. 2016. "High Potential for Weathering and Climate Effects of Non-vascular Vegetation in the Late Ordovician." *Nature Communications* 7: 12113.

Powers, S., and L. Lehmann. 2013. "The Co-Evolution of Social Institutions, Demography, and Large-Scale Human Cooperation." *Ecology Letters* 16(11): 1356–64.

———. 2017. "When Is Bigger Better? The Effects of Group Size on the Evolution of Helping Behaviours." *Biological Reviews* 92(2): 902–20.

Powers, S., C. Van Schaik, and L. Lehmann. 2016. "How Institutions Shaped the Last Major Evolutionary Transition to Large-Scale Human Societies." *Philosophical Transactions of the Royal Society B* 371(1687): 20150098.

Pradeu, T., ed. 2016. "Biological Individuality." Special issue, *Biology and Philosophy* 31(6).

Priest, G. 2014. *One: Being an Investigation into the Unity of Reality and of Its Parts, Including the Singular Object Which Is Nothingness*. Oxford: Oxford University Press.

Pyne, S. 2016. "Fire in the Mind: Changing Understandings of Fire in Western Civilization." *Philosophical Transactions of the Royal Society B* 371(1696): 20150166.

Rahn, H., and T. Yokoyama, eds. 1965. *Physiology of Breath-Hold Diving and the Ama of Japan*. Washington, DC: National Academies of Science.

Read, D. 2006. "Tasmanian Knowledge and Skill: Maladaptive Imitation or Adequate Technology?" *American Antiquity* 71(1): 164–84.

———. 2008. "An Interaction Model for Resource Implement Complexity Based on Risk and Number of Annual Moves." *American Antiquity* 73(4): 599–625.

Reichhold, J., trans. 2008. *Basho: The Complete Haiku*. Tokyo: Kodansha.

Reynolds, H. 2012. *A History of Tasmania*. Cambridge: Cambridge University Press.

Rhodes, E. 2011. "Optically Stimulated Luminescence Dating of Sediments over the Past 200,000 Years." *Annual Review of Earth and Planetary Sciences* 39: 461–88.

Roebroeks, W., and P. Villa. 2011. "On the Earliest Evidence for Habitual Use of Fire in Europe." *Proceedings of the National Academy of Sciences of the United States of America* 108(13): 5209–14.

Ryan, L. 2013. "The Black Line in Van Diemen's Land (Tasmania), 1830." *Journal of Australian Studies* 37(1): 1–8.

Rusch, H. 2014. "The Evolutionary Interplay of Intergroup Conflict and Altruism in Humans: A Review of Parochial Altruism Theory and Prospects for Its Extension." *Proceedings of the Royal Society B* 281(1794): 20141539.

Samson, D., A. Crittenden, I. Mabulla, A. Mabulla, and C. Nunn. 2017. "Hadza Sleep Biology: Evidence for Flexible Sleep-Wake Patterns in Hunter-Gatherers." *American Journal of Physical Anthropology* 162(3): 573–82.

Samson, D., and C. Nunn. 2015. "Sleep Intensity and the Evolution of Human Cognition." *Evolutionary Anthropology* 24(6): 225–37.

Sandgathe, D. 2017. "Identifying and Describing Pattern and Process in the Evolution of Hominin Use of Fire." *Current Anthropology* 58(S16): S360–70.

Sandgathe, D., and F. Berna, eds. 2017. "Fire and Genus *Homo*." Special issue, *Current Anthropology* 58(S16).

Sasaki, F. 2017. *Goodbye, Things: The New Japanese Minimalism*. Translated by E. Sugita, E. New York: Norton.

Schwartz, G. 2012. "Growth, Development, and Life History throughout the Evolution of *Homo*." *Current Anthropology* 53(S6): S395–408.

Scott, A., W. Chaloner, C. Belcher, and C. Roos, eds. 2016. "The Interaction of Fire and Mankind." Special issue, *Philosophical Transactions of the Royal Society B* 371(1696).

Scott, J. 2009. *The Art of Not Being Governed: An Anarchist History of Upland Southeast Asia*. New Haven, CT: Yale.

Shennan, S. 2001. "Demography and Cultural Innovation: A Model and Its Implications for the Emergence of Modern Human Culture." *Cambridge Archaeological Journal* 11(1): 5–16.

Skyrms, B. 2010. *Signals: Evolution, Learning, and Information.* Oxford: Oxford University Press.

Smith, M., and S. Myers. 2018. "Impact of Anthropogenic CO_2 Emissions on Global Human Nutrition." *Nature Climate Change* 8: 834–39.

Snodgrass, J., M. Sorensen, L. Tarskala, and W. Leonard. 2007. "Adaptive Dimensions of Health Research among Indigenous Siberians." *American Journal of Human Biology* 19(1): 165–80.

Steffen, W., J. Rockström, K. Richardson, T. Lenton, C. Folke, et al. 2018. "Trajectories of the Earth System in the Anthropocene." *Proceedings of the National Academy of Sciences of the United States of America* 115(33): 8252–59.

Stepniak, B., S. Papiol, C. Hammer, A. Ramin, S. Everts, et al. 2014. "Accumulated Environmental Risk Determining Age at Schizophrenia Onset: A Deep Phenotyping-Based Study." *Lancet Psychiatry* 1(6): 444–53.

Sterelny, K. 2011. "From Hominins to Humans: How *Sapiens* Became Behaviourally Modern." *Philosophical Transactions of the Royal Society* B 366(1566): 809–22.

Stewart, J., O. Gapenne, E. Di Paolo, eds. 2010. *Enaction: Toward a New Paradigm for Cognitive Science.* Cambridge, MA: MIT Press.

Stilgoe, J. 2016. *What Is Landscape?* Cambridge, MA: MIT Press.

Taylor, R. 2014. "Archaeology and Aboriginal Protest." *Australian Historical Studies* 45(3): 331–49.

Terashima, H., and B. Hewlett, eds. 2018. *Social Learning and Innovation in Contemporary Hunter-Gatherers: Evolutionary and Ethnographic Perspectives.* Tokyo: Springer.

Thompson, E. 2007. *Mind in Life: Biology, Phenomenology, and the Sciences of Mind.* Cambridge, MA: Harvard University Press.

Tobler, R., A. Rohrlach, J. Soubrier, P. Bover, B. Llamas, et al. 2017. "Aboriginal Mitogenomes Reveal 50,000 Years of Regionalism in Australia." *Nature* 544(7649): 180–84.

Tomasello, M. 2008. *Origins of Human Communication.* Cambridge, MA: MIT Press.

Turing, A. 1952. "The Chemical Basis of Morphogenesis." *Philosophical Transactions of the Royal Society* B 237(1): 37–72.

United Nations High Commissioner for Refugees. 2018. *Global Trends: Forced Displacement in 2017.* Geneva: UNHCR.

Vaesen, K., M. Collard, R. Cosgrove, and W. Roebroeks. 2016 "Population Size Does Not Explain Past Changes in Cultural Complexity." *Proceedings of the National Academy of Sciences of the United States of America* 113(16): E2241–47.

Van der Peer, Y., E. Mizrachi, and K. Marchal. 2015. "The Evolutionary Significance of Polyploidy." *Nature Reviews Genetics* 18: 411–24.

Vitousek, S., P. Barnard, C. Fletcher, N. Frazer, L. Erikson, and C. Storlazzi. 2017. "Doubling of Coastal Flooding Frequency within Decades Due to Sea-Level Rise." *Scientific Reports* 7: 1399.

Wanner, H., J. Beer, J. Bütikofer, T. Crowley, U. Cubasch, et al. 2008. "Mid- to Late Holocene Climate Change: An Overview." *Quaternary Science Reviews* 27(19–20): 1791–1828.

Wehr, T. 2001. "Photoperiodism in Humans and Other Primates: Evidence and Implications." *Journal of Biological Rhythms* 16(4): 348–64.

Weiner, A. 1994. "Cultural Difference and the Density of Objects." *American Ethnologist* 21(2): 391–403.

Wiedenhofer, D., B. Smetschka, L. Akenji, M. Jalas, and H. Haberl. 2018. "Household Time Use, Carbon Footprints, and Urban Form: A Review of the Potential Contributions of Everyday Living to the 1.5° C Climate Target." *Current Opinion in Environmental Sustainability* 30(1): 7–17.

Williams, J., Y. Masuda, and H. Tallis. 2016. "A Measure Whose Time Has Come: Formalizing Time Poverty." *Social Research Indicators* 128(1): 265–83.

Willis, K., and J. McElwain. 2014. *The Evolution of Plants,* 2nd ed. Oxford: Oxford University Press.

Wimsatt, W. 2014. "Entrenchment and Scaffolding: An Architecture for a Theory of Cultural Change." In *Developing Scaffolds in Evolution, Culture, and Cognition,* edited by L. Caporeal, J. Griesemer, and W. Wimsatt, 77–105. Cambridge, MA: MIT Press.

Worm, B., H. Lotze, I. Jubinville, C. Wilcox, J. Jambeck. 2017. "Plastic as a Persistent Marine Pollutant." *Annual Review of Environment and Resources* 42(1): 1–26.

Worthman, C. 2011. "Developmental Cultural Ecology of Sleep." In *Sleep and Development: Familial and Socio-Cultural Considerations,* edited by M. El-Sheikh, 167–94. New York: Oxford University Press.

Wright, S. 1932. "The Role of Mutation, Inbreeding, Crossbreeding and Selection in Evolution." In *Proceedings of the Sixth International*

Congress of Genetics, Ithaca, New York, vol. 1, *Transactions and General Addresses*, edited by D. Jones, 356–66. Brooklyn: Brooklyn Botanic Garden.

Wynn, T., and F. Coolidge, eds. 2010. "Working Memory." *Current Anthropology* 51 (S1).

Yu, P., R. Xu, M. Abramson, S. Li, and Y. Guo. 2020. "Bushfires in Australia: A Serious Health Emergency under Climate Change." *Lancet Planetary Health* 4 (1): e7–8.

Zelizer, V. 2007. *The Purchase of Intimacy.* Princeton, NJ: Princeton University Press.

Index

accelerated weathering, 153–54
adaptation, 50–54; vs. acclimatization,
90; climate change and, 96, 154; multiple time horizons of, 56–57, 90–91,
96, 151, 158–59, 160*fig. See also* cold,
adaptation to; landscapes, adaptive;
scaffolds
ama (Japanese female breath-hold
divers), 48–51, 164*fig.*
Anthropological Expedition, Cambridge,
28, 29*fig.*
apnea diving. *See* diving, breath-hold
asceticism, xxvi, 81, 127

bark-bundle canoes, 18, 19*fig.*, 25, 37, 63,
164*fig.*
Basho (poet, early Edo Japan), 132
BAT. *See* brown adipose tissue
Behavioral Risk Factor Surveillance System (US Centers for Disease Control
and Prevention), 92–96
boro (textile patching technique), xxvi–
xxviii, 26, 98, 110, 133, 169
breath-hold diving. *See* diving, breath-hold

brown adipose tissue, 23, 50, 54, 164*fig.*,
178n23
bushfire, xxix–xxxi, 25, 91. *See also*
wildfire

Cambridge Anthropological Expedition to
Torres Straits, 28, 29*fig.*
canoes, bark-bundle, 18, 19*fig.*, 25, 37, 63,
164*fig.*
chaîne opératoire, 37, 148, 171
climate change, implications for behavior,
91–98
clothing, archaeology of, 20–22
cold, adaptation to, 31, 68, 73, 78, 91,
178n23; in Tasmania, 10, 19–24, 66.
See also brown adipose tissue; diving,
breath-hold
complexity, social. *See* social complexity
costly signaling, 32. *See also* skills
reservoir
Covid-19, 94, 146
CP 1919 (pulsar), 74*fig.*, 75
cumulativity, 5, 91, 147–50, 159, 161–62,
169. *See also* treadmills

decluttering, 130–32, 138
development, economic and social-structural, archaeology of, 59, 65, 71. *See also* intensification, archaeology of
displaced and unhoused people, 129
ditch kit, 126–28
diving, breath-hold, 46–52, 54, 57, 68, 164*fig.*; physiological effects of, 50–51; in Tasmania, 47. *See also* brown adipose tissue
Dunbar's number, 103–4. *See also* social complexity

El Niño Southern Oscillation (ENSO), 61, 142
enaction, xx, xxii–xxv, 54, 112, 122, 148–49, 154, 169
enregisterment, 40–41, 53, 100, 169; of currencies, 136; of institutions, 106
epiphytes, xv–xvii, xxvi, 133. *See also* ditch kit
evolutionary game theory. *See* game theory
exaptation, 68, 73, 112
exercise. *See* physical activity, trends in
extrinsic risk, 17, 33, 161, 171. *See also* skills reservoir

fire: archaeology of, 25–26, 63, 72–73; curation of, 25–26, 31, 37; emotional resonance of, 56; use in niche construction, 8–9, 34, 57–58, 160*fig.*, 164*fig.*; in one's peripersonal space, 72; skillfulness vs. technological complexity, 31. *See also* bushfire; Southwest Australia; wildfire
flourishing, 48–54, 96–97, 102, 111–12, 159
foaminess, 166–167, 170. *See also* reaction-diffusion process; social complexity
forcings, extrinsic (climate), 60–61, 80, 90, 142, 147

games with stuff, 133–40, 141*fig.*
game theory, xv, 161, 170, 173–74, 178n18a. *See also* strategy
Gray-Scott model. *See* reaction-diffusion process
Great Pacific Garbage Patch. *See* plastics, accumulation in the biosphere

group selection, 102, 104. *See also* inclusive fitness; social complexity; ultrasociality

haenyeo (Korean female breath-hold divers), 48–51, 50*fig.*, 57. *See also* ama; diving, breath-hold
Hobbesian war of all against all, 107. *See also* social complexity

image schemata, 113, 118, 171
inclusive fitness, 101–3. *See also* group selection; social complexity; ultrasociality
individuality, biological, 107–8*fig. See also* foaminess; reaction-diffusion process; social complexity
innovation: design profession and, xviii; discounting of enactive innovation, 46–47, 122; patents as a measure of, 3; in social structure, 64–65. *See also* cumulativity; learning, social or observational; strategy
intensification, archaeology of, 59, 64–65, 70–71

Jones, Rhys (archaeologist), 12–15, 34, 67, 69–70, 72, 178n16a
Joy Division (cover art for *Unknown Pleasures*), 74*fig.*, 75

Kondo, Marie (decluttering guru), 129–32

landscapes, adaptive, 79–80, 110–11; category theory and, 118–19; divergent topologies of, 81–84, 114–17; semantic templates and, 112–13
learning, social or observational, 35–36, 41–47, 102, 147
Life-Changing Magic of Tidying Up, The (self-help book by Marie Kondo), 129–30
living epiphytically. *See* epiphytes

Malinowski, Bronisław (anthropologist), xxi, 134
mass mortality, 146
meditation, xii, xxiii, 77, 126
megafauna extinctions, 145–46

Million Random Digits with 100,000 Normal Deviates, A, 128
minimalism. *See* decluttering

niche construction, xv, 3, 38, 80, 99, 164*fig.*; vs. adaptation, 53, 83; use of fire in, 34, 68, 90; somatic, xxiii, 69, 122; and the topology of evolution, 75, 118, 183n80
Noongar (Indigenous Southwest Australians), 64, 72, 167

operatory chain, 37, 148, 171

physical activity, trends in, 94–96
planning, archaeology of, 67, 131–32. *See also* adaptation; scale, methodological challenge of
plastics, accumulation in the biosphere, 140–44
population games, 160*fig. See also* game theory; strategy
popup studio. *See* Cambridge Anthropological Expedition to Torres Straits

reaction-diffusion process, 107, 108*fig.*, 165. *See also* social complexity
refugees, 129
registers, theory of. *See* enregisterment
risk, extrinsic, 17, 33, 161, 171. *See also* skills reservoir
Rivers, W. H. R. (anthropologist), xxi, *xxix*
The Road (novel by Cormac McCarthy), 78, 151–52
Robinson, George Augustus (colonial ethnographer), 13–15, 25, 66

saddles (in adaptive landscapes), 59, 67
SARS-CoV-2. *See* COVID-19
scaffolds: artifacts as, 38–39; bodies as, 33; vs. landscapes, 120–24, 123*fig.*; in tissue fabrication, 120–21. *See also* landscapes, adaptive
scale, methodological challenge of, 26–29, 35, 69, 74, 98–107, 172
semantic templates, 112–13, 172
semiokinetic ecology, xviii, xx, 163–66, 172
skills reservoir, 18, 147–48, 151, 158–63, 160*fig.*, 173, 178n18a

sleep, plasticity of, 96–98
social acceleration, 127, 154
social complexity, 98–107
Southwest Australia: climate and biota of, 59–63; Pleistocene colonization of, 63–65. *See also* Noongar
strategy (*sensu* behavioral ecology, game theory), 173; cooperative hunting in, 45; cumulativity and, 147; decision theory and, 163; long-term equilibria in, 65–66; of fire, 25, 70–74; flourishing and, 114; gender and, 17–18, 31–32; inferring from artifacts, 54; planning and, 68; prey-rank models and, 64; of social learning, 41–43; windfall strategies, 17–18. *See also* cumulativity; flourishing; strictly dominated strategy
strictly dominated strategy, 160*fig.*, 161, 178n18a. *See also* skills reservoir
stuff, anxiety about, 129, 132. *See also* decluttering
stuff, games with, 133–40, 141*fig.*

Tamarisk Row (novel by Gerald Murnane), vi, 28–29, 136–38, 141*fig.*
Tasmania: climate and biota of, 6–9; Pleistocene colonization of, 9–12
Tasmanians, Indigenous: clothing, 18–23; colonization and dispossession, 13–15, 177n14a; diet and nutrition, 11, 16–18, 34–35; fire, use of, 24–26; hunting strategies, 17–18. *See also* cold, adaptation to; thermoregulation
templates, semantic, 112–13, 172
thermogenesis, 23, 50, 54, 164*fig.*, 178n23. *See also* brown adipose tissue
thermoregulation, 31, 50. *See also* brown adipose tissue; cold, adaptation to; thermogenesis
time, multiple horizons of, 56–57, 90–91, 96, 151, 158–59, 160*fig. See also* adaptation
treadmills, 1–6, 30–38, 40–44, 148–49, 174. *See also* cumulativity
trophic truncation, 144–46

ultrasociality, 98–99. *See also* social complexity

urbanization, stressors of, 85–91; aerosol pollution, 85–86; noise, 87–88; and oxidative stress, 89–90; and schizophrenia, 88–89

view from nowhere, the (*sensu* Thomas Nagel), 121

weathering, accelerated, 153–54
wildfire, 25–26, 79, 81, 145, 189n147b. *See also* bushfire

zazen (sitting meditation), xii, xxiii, 77
zoonoses, 145–46

Founded in 1893,
UNIVERSITY OF CALIFORNIA PRESS
publishes bold, progressive books and journals
on topics in the arts, humanities, social sciences,
and natural sciences—with a focus on social
justice issues—that inspire thought and action
among readers worldwide.

The UC PRESS FOUNDATION
raises funds to uphold the press's vital role
as an independent, nonprofit publisher, and
receives philanthropic support from a wide
range of individuals and institutions—and from
committed readers like you. To learn more, visit
ucpress.edu/supportus.